高等学校电气工程与自动化专业系列教材

电气工程专业英语实用教程
第4版

张强华 司爱侠 ◎ 编著

清华大学出版社
北京

内 容 简 介

本书的目的在于切实提高读者的专业英语能力。

本书体例上以 Unit 为单位，每一 Unit 由以下几部分组成：课文——这些课文包括了基础知识和基本概念；单词、词组及缩略语——给出课文中出现的新词、常用词组及缩略语，读者由此可以积累电气专业的基本词汇，单词均加注了音标；难句讲解——讲解课文中出现的疑难句子，培养读者的阅读理解能力；习题——既有针对课文的练习，也有一些开放性的练习；科技英语翻译知识——帮助读者掌握基本的专业英语翻译技巧；阅读材料——提供最新的设备和工具软件的相关资料，可进一步扩大读者的视野。书末还提供了自测题及参考答案，可供读者检查学习效果。

本书既可作为高等学校电气工程专业英语教材，也可作为培训班教材和供从业人员自学。

版权所有，侵权必究。举报：010-62782989，beiqinquan@tup.tsinghua.edu.cn。

图书在版编目（CIP）数据

电气工程专业英语实用教程 / 张强华，司爱侠编著.
4 版. -- 北京 : 清华大学出版社, 2024.7（2025.2重印）. -- (高等学校电气工程与自动化专业系列教材). -- ISBN 978-7-302-66738-4

Ⅰ. TM

中国国家版本馆CIP数据核字第2024LU9155号

责任编辑：安　妮
封面设计：刘　键
责任校对：郝美丽
责任印制：刘　菲

出版发行：清华大学出版社
网　　址：https://www.tup.com.cn, https://www.wqxuetang.com
地　　址：北京清华大学学研大厦 A 座　　邮　编：100084
社 总 机：010-83470000　　邮　购：010-62786544
投稿与读者服务：010-62776969, c-service@tup.tsinghua.edu.cn
质 量 反 馈：010-62772015, zhiliang@tup.tsinghua.edu.cn
课 件 下 载：https://www.tup.com.cn, 010-83470236
印 装 者：三河市君旺印务有限公司
经　　销：全国新华书店
开　　本：185mm×260mm　印　张：16.5　字　数：402 千字
版　　次：2016 年 1 月第 1 版　2024 年 8 月第 4 版　印　次：2025 年 2 月第 2 次印刷
印　　数：1501～2700
定　　价：59.00 元

产品编号：104675-01

前 言

现如今,电气行业的新技术、新设备、新工具不断出现,要掌握这些新知识和新技能,从业人员就必须不断地学习,提高专业英语水平。为此,就必须进行针对性的专门学习。本书的目的就在于切实提高读者实际使用电气专业英语的能力。

本书课文包括电的基本概念、交流电简介、电阻和电导、电容、简单电路、直流并联电路、微控制器、集成电路、模拟电路和数字电路、数字电路元件、电路分析和设计、直流电动机、控制概论、数字控制系统、电子设计自动化、SPICE 软件、可编程逻辑控制器、智能电网、人工智能在电气工程中的应用及电气工程师的主要技能。本书阅读材料包括电气工程简介、电感、OrCAD 软件、电气 AutoCAD、嵌入式系统、电网、物联网、电气工程常用软件、VHDL 软件及可再生能源。

本书体例上以 Unit 为单位,每一 Unit 由以下几部分组成:课文——内容包括了基础知识和基本概念;单词、词组及缩略语——给出课文中出现的新词、常用词组及缩略语,读者由此可以积累电气专业方面的基本词汇,单词均加注了音标;难句讲解——讲解课文中出现的疑难句子,培养读者的阅读理解能力;习题——既有针对课文的练习,也有一些开放性的练习;科技英语翻译知识——帮助读者掌握基本的专业英语翻译技巧;阅读材料——提供最新的设备和工具软件的相关资料,可进一步扩大读者的视野。书末还提供了自测题及参考答案,可供读者检查学习效果。

本书适应当前许多院校已经实行的"宽口径"人才培养模式,新增了人工智能、智能电网及可再生能源等新内容。

本书在编写中,着重从"教师教什么""学生就业后用什么"来考虑并结合学生的具体情况,针对学生毕业后的就业环境,根据未来工作实际的要求,做了切合实际的精心加工。本书提供教学课件、教学大纲、习题答案、模拟试卷及全书总词汇表。

望大家不吝赐教,使本书成为一部"符合学生实际、切合行业实况、知识实用丰富、严谨开放创新"的优秀教材。

本书既可作为高等院校电气工程专业英语教材,也可作为培训班教材和供从业人员自学。

编 者

2024 年 3 月

目 录

Unit 1 ··· 1

  Text A   Basic Concepts of Electricity ··· 1
      New Words ··· 5
      Phrases ··· 6
      Notes ··· 6
      Exercises ··· 7
  Text B   Introduction to AC ··· 9
      New Words ··· 14
      Phrases ··· 15
      Abbreviations ·· 15
      Exercises ·· 15
  科技英语翻译知识   翻译的标准 ·· 16
      Reading Material：Introduction to Electrical Engineering ····························· 17
  参考译文   电的基本概念 ·· 21

Unit 2 ·· 23

  Text A   Electrical Resistance and Conductance ··· 23
      New Words ··· 26
      Phrases ··· 27
      Abbreviations ·· 28
      Notes ·· 28
      Exercises ·· 29
  Text B   Capacitor ·· 31
      New Words ··· 35
      Phrases ··· 36
      Abbreviations ·· 37
      Exercises ·· 37
  科技英语翻译知识   词义的选择 ·· 37
      Reading Material：Inductor ··· 39
  参考译文   电阻和电导 ·· 44

Unit 3 ... 46

Text A　Simple Electric Circuit ... 46
New Words ... 50
Phrases ... 51
Notes ... 52
Exercises ... 52

Text B　DC Parallel Circuit ... 55
New Words ... 60
Phrases ... 61
Exercises ... 61

科技英语翻译知识　词义的引申 ... 63
Reading Material：OrCAD View ... 64

参考译文　简单电路 ... 70

Unit 4 ... 73

Text A　Microcontroller ... 73
New Words ... 76
Phrases ... 77
Abbreviations ... 78
Notes ... 79
Exercises ... 80

Text B　Integrated Circuits ... 82
New Words ... 86
Phrases ... 87
Abbreviations ... 88

科技英语翻译知识　词义的增减 ... 90
Reading Material：Introduction to AutoCAD Electrical ... 92

参考译文　微控制器 ... 96

Unit 5 ... 99

Text A　Analog Circuits and Digital Circuits ... 99
New Words ... 103
Phrases ... 104
Abbreviations ... 105
Notes ... 105
Exercises ... 106

Text B　Digital Circuit Elements ... 108
New Words ... 111

- Phrases ·· 112
  - Abbreviations ·· 113
  - Exercises ·· 113
- 科技英语翻译知识　词类的转换 ································ 113
  - Reading Material：Embedded System ····················· 115
- 参考译文　模拟电路和数字电路 ································ 118

Unit 6 ·· 122

- Text A　Electronic Circuit Analysis and Design ············· 122
  - New Words ·· 125
  - Phrases ·· 126
  - Abbreviations ·· 127
  - Notes ·· 127
  - Exercises ·· 128
- Text B　Basic DC Motor Operation ··························· 131
  - New Words ·· 134
  - Phrases ·· 135
  - Abbreviations ·· 135
  - Exercises ·· 135
- 科技英语翻译知识　否定的译法 ································ 136
  - Reading Material：Power Grid ······························ 138
- 参考译文　电路分析与设计 ······································ 141

Unit 7 ·· 145

- Text A　Basic Principles of Control ··························· 145
  - New Words ·· 147
  - Phrases ·· 148
  - Notes ·· 149
  - Exercises ·· 150
- Text B　Digital Control Systems ································ 153
  - New Words ·· 156
  - Phrases ·· 157
  - Abbreviations ·· 157
  - Exercises ·· 157
- 科技英语翻译知识　被动语态的译法 ························· 158
  - Reading Material：Internet of Things ····················· 160
- 参考译文　控制的基本原理 ······································ 165

Unit 8 ... 167

Text A Electronic Design Automation ... 167
- New Words ... 171
- Phrases ... 172
- Abbreviations ... 172
- Notes ... 173
- Exercises ... 174

Text B SPICE ... 177
- New Words ... 180
- Phrases ... 181
- Abbreviations ... 182
- Exercises ... 182

科技英语翻译知识 从句的译法 ... 183

Reading Material：Electrical Engineering Software ... 186

参考译文 电子设计自动化 ... 190

Unit 9 ... 193

Text A Programmable Logic Controllers (PLCs) ... 193
- New Words ... 198
- Phrases ... 199
- Abbreviations ... 200
- Notes ... 200
- Exercises ... 201

Text B Smart Power Grids ... 204
- New Words ... 207
- Phrases ... 208
- Exercises ... 208

科技英语翻译知识 汉语四字格的运用 ... 209

Reading Material：VHDL ... 211

参考译文 可编程逻辑控制器（PLC） ... 218

Unit 10 ... 222

Text A Applications of Artificial Intelligence in Electrical Engineering ... 222
- New Words ... 225
- Phrases ... 227
- Abbreviations ... 228
- Notes ... 228
- Exercises ... 229

Text B	Top Skills for an Electrical Engineer	232
	New Words	236
	Phrases	237
	Exercises	237

科技英语翻译知识　篇章翻译 ······ 238
　　Reading Material：Types of Renewable Energy Resources ······ 239
参考译文　人工智能在电气工程中的应用 ······ 243

附录 A　自测题及参考答案 ······ 246

Text A

Basic Concepts of Electricity

1. Electric Charges

1.1 Neutral state of an atom

Elements are often identified by the number of electrons in orbit around the nucleus of the atoms making up the element and by the number of protons in the nucleus.[1] A hydrogen atom, for example, has only one electron and one proton. An aluminum atom has 13 electrons and 13 protons (see Figure 1-1). An atom with an equal number of electrons and protons is said to be electrically neutral.

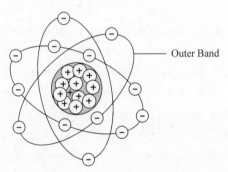

Figure 1-1   An aluminum atom

1.2 Positive and negative charges

Electrons in the outer band of an atom are easily displaced by the application of some external force. Electrons which are forced out of their orbits can result in a lack of electrons where they leave and an excess of electrons where they come to rest.[2] The lack of electrons is called a positive charge because there are more protons than electrons. The excess of electrons has a negative charge. A positive or negative charge is caused by an absence or excess of electrons. The number of protons remains constant (see Figure 1-2).

1.3 Attraction and repulsion of electric charges

The old saying "opposites attract" is true when dealing with electric charges. Charged bodies have an invisible electric field around them. When two like-charged bodies are brought

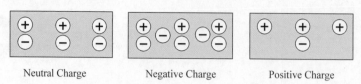

Figure 1-2　Positive and negative charges

together, their electric field will work to repel them. When two unlike-charged bodies are brought together, their electric field will work to attract them. The electric field around a charged body is represented by invisible lines of force. The invisible lines of force represent an invisible electrical field that causes the attraction and repulsion. Lines of force are shown leaving a body with a positive charge and entering a body with a negative charge (see Figure 1-3).

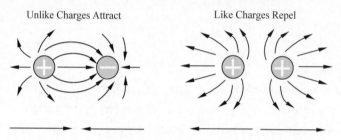

Figure 1-3　Attraction and repulsion of electric charges

2. Current

Electricity is the flow of free electrons in a conductor from one atom to the next atom in the same general direction. This flow of electrons is referred to as current and is designated by the symbol "I". Electrons move through a conductor at different rates and electric current has different values. Current is determined by the number of electrons that pass through a cross-section of a conductor in one second. We must remember that atoms are very small. It takes about 1,000,000,000,000,000,000,000,000 atoms to fill one cubic centimeter of a copper conductor. This number can be simplified using mathematical exponents. Instead of writing 24 zeros after the number 1, write 10^{24}. Trying to measure even small values of current would result in unimaginably large numbers.[3] For this reason current is measured in amperes which is abbreviated "amp". The symbol for amp is the letter "A". A current of one amp means that in one second about 6.24×10^{18} electrons move through a cross-section of conductor. These numbers are given for information only and you do not need to be concerned with them. It is important, however, that the concept of current flow be understood (see Figure 1-4).

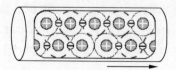

Figure 1-4　Current flow

2.1 Units of measurement for current

The following Table 1-1 reflects special prefixes that are used when dealing with very small or large values of current:

Table 1-1 Units of measurement for current

Prefix	Symbol	Decimal
1 kiloampere	1kA	1000A
1 milliampere	1mA	1/1000A
1 microampere	1μA	1/1 000 000A

2.2 Direction of current flow

Some authorities distinguish between electron flow and current flow. Conventional current flow theory ignores the flow of electrons and states that current flows from positive to negative. To avoid confusion, we will use the electron flow concept which states that electrons flow from negative to positive (see Figure 1-5).

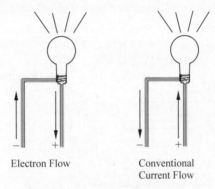

Figure 1-5 Direction of current flow

3. Voltage

Electricity can be compared with water flowing through a pipe (see Figure 1-6). A force is required to get water to flow through a pipe. This force comes from either a water pump or gravity. Voltage is the force that is applied to a conductor that causes electric current to flow. Electrons are negative and are attracted by positive charges. They will always be attracted from a source having an excess of electrons, thus having a negative charge, to a source having a deficiency of electrons which has a positive charge.[4] The force required to make electricity flow through a conductor is called a difference in potential, electromotive force (emf), or more simply referred to as voltage. Voltage is designated by the letter "U", or the letter "V". The unit of measurement for voltage is volt which is designated by the letter "V".

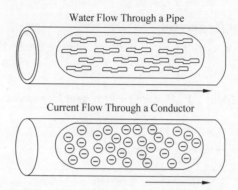

Figure 1-6　Current flow through a conductor

3.1　Voltage sources

An electric voltage can be generated in various ways. A battery uses an electrochemical process. A car's alternator and a power plant generator utilize a magnetic induction process. All voltage sources share the characteristic of an excess of electrons at one terminal and a shortage at the other terminal. This results in a difference of potential between the two terminals (see Figure 1-7).

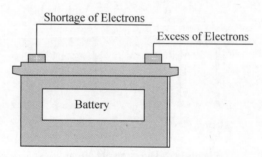

Figure 1-7　Battery

3.2　Voltage symbol

The terminals of a battery are indicated symbolically on an electrical drawing by two lines. The longer line indicates the positive terminal. The shorter line indicates the negative (see Figure 1-8).

Figure 1-8　Voltage symbol

3.3　Units of measurement for voltage

The following Table 1-2 reflects special prefixes that are used when dealing with very small or large values of voltage:

Table 1-2　Units of measurement for voltage

Prefix	Symbol	Decimal
1 kilovolt	1kV	1000V
1 millivolt	1mV	1/1000V
1 microvolt	1μV	1/1 000 000V

New Words

atom	['ætəm]	n. 原子
orbit	['ɔːbɪt]	n. 轨道；轨迹
designate	['dezɪɡneɪt]	vt. 指定(出示)，标明
nucleus	['njuːklɪəs]	n. 核子
cubic	['kjuːbɪk]	adj. 立方体的，立方的
proton	['prəʊtɒn]	n. 质子
abbreviate	[ə'briːvɪeɪt]	vt. 节略，省略，缩写
excess	['ekses]	n. 过度，多余，超过，超额
		adj. 过度的，额外的
conventional	[kən'venʃənl]	adj. 惯例的，常规的，习俗的，传统的
gravity	['ɡrævətɪ]	n. 地心引力，重力
charge	[tʃɑːdʒ]	n. 负荷，电荷，费用，充电
absence	['æbsəns]	n. 缺乏，没有
remain	[rɪ'meɪn]	vi. 剩余，残存
attraction	[ə'trækʃn]	n. 吸引，吸引力
repulsion	[rɪ'pʌlʃn]	n. 排斥
repel	[rɪ'pel]	vt. 排斥
represent	[ˌreprɪ'zent]	vt. 表示，表现
symbol	['sɪmbl]	n. 符号，记号
rate	[reɪt]	n. 比率，速度，等级
determine	[dɪ'tɜːmɪn]	vt. & vi. 决定，确定，测定
cross-section	['krɒs sekʃn]	n. 截面，断面
exponent	[ɪk'spəʊnənt]	n. 指数，幂
unimaginably	['ʌnɪ'mædʒɪnəblɪ]	adj. 不能想象的，难以理解的
measurement	['meʒəmənt]	n. 量度，测量法
prefix	['priːfɪks]	n. 前缀
distinguish	[dɪ'stɪŋɡwɪʃ]	vt. & vi. 区别，辨别
ignore	[ɪɡ'nɔː]	vt. 忽略，不理睬，忽视
electrochemical	[ɪˌlektrəʊ'kemɪkəl]	adj. 电气化学的
alternator	['ɔːltəneɪtə]	n. 交流发电机
generator	['dʒenəreɪtə]	n. 发电机

magnetic	[mæg'netɪk]	adj. 磁的，有磁性的，有吸引力的
deficiency	[dɪ'fɪʃnsɪ]	n. 缺乏，不足
potential	[pə'tenʃl]	adj. 势的，位的
utilize	['juːtəlaɪz]	vt. 利用
terminal	['tɜːmɪnl]	n. 终端，接线端，电路接头
characteristic	[ˌkærəktə'rɪstɪk]	adj. 特有的，典型的
		n. 特性，特征
shortage	['ʃɔːtɪdʒ]	n. 不足，缺乏
symbolically	[sɪm'bɒlɪklɪ]	adv. 象征性地
kilovolt	['kɪləvəʊlt]	n. 千伏特
millivolt	['mɪlɪvəʊlt]	n. 毫伏（特）[=1/1000 伏（特），略作 mV]
microvolt	['maɪkrəʊvəʊlt]	n. 微伏[等于 1 伏（特）的百万分之一]

Phrases

neutral state	中性状态
force out	挤（出去），冲（出去）
negative charge	负电荷
positive charge	正电荷
electric field	电场
free electron	自由电子
be concerned with	关心
current flow	电流

Notes

[1] Elements are often identified by the number of electrons in orbit around the nucleus of the atoms making up the element and by the number of protons in the nucleus.

本句中，and 连接了两个由介词 by 引导的方式状语。在第一个方式状语中，making up the element 是一个现在分词短语，它作定语，修饰和限定 atoms；in orbit around the nucleus of the atoms 是一个介词短语，它作定语，修饰和限定 electrons。在第二个方式状语中，介词短语 in the nucleus 作定语，修饰和限定 protons。be identified by 的意思是"通过……鉴别""识别"。the number of 的意思是"……的数量"。

本句意为：元素通常通过组成该元素的原子核周围轨道上的电子数量和原子核中的质子数量来识别。

[2] Electrons which are forced out of their orbits can result in a lack of electrons where they leave and an excess of electrons where they come to rest.

本句中的主语是 electrons，谓语是 can result in，a lack of electrons where they leave and an excess of electrons where they come to rest 是宾语。这个宾语由 and 连接的名词短语 a lack of electrons 及 an excess of electrons 组成，两个 where 引导地点状语，which are forced out of their orbits 是一个定语从句，修饰和限定 Electrons，在理解时要分清这里面的层次。

本句意为：电子从轨道移走会造成该原子缺少电子，而移入电子的原子会使电子过剩。

[3] Trying to measure even small values of current would result in unimaginably large numbers.

本句的主语是一个动名词短语 trying to measure even small values of current，谓语是动词词组 result in，result in 不能机械地理解为"导致"，因为"导致"带有明显的感情色彩，理解为"产生"更顺畅。

本句意为：试图测量即使是很小的电流也将产生难以想象的巨大的数字。

[4] They will always be attracted from a source having an excess of electrons, thus having a negative charge, to a source having a deficiency of electrons which has a positive charge.

本句的主干是 they will always be attracted from…to…。having an excess of the electrons, thus having a negative charge 两个定语修饰的是第一个 source。having a deficiency of electrons which has a positive charge 修饰的是第二个 source。第二个修饰语虽然没有明显表达出两个定语之间的逻辑关系，但是从第一个修饰语中两个定语用 thus 连接可以判断出，它们仍然有因果关系：因为缺乏电子，才形成正电荷。所以在理解时要注意这层意思。

本句意为：电子总是从由于多余电子而呈现负电荷的地方被吸引到由于缺乏电子而呈现正电荷的地方。

Exercises

【Ex.1】根据课文内容，回答以下问题。

1. What is a hydrogen atom made up of ?

2. Why does an atom present neutral?

3. How does an atom have positive charge?

4. What will happen when two like-charged bodies are brought together?

5. What is defined about the current according to conventional current flow theory?

【Ex.2】根据下面的英文解释，写出相应的英文词汇。

英 文 解 释	词　汇
a tiny particle of matter that is smaller than an atom and has a negative electrical charge	

续表

英 文 解 释	词　汇
a colorless, highly flammable gaseous element, the lightest of all gases and the most abundant element in the universe	
the net measure of this property possessed by a body or contained in a bounded region of space	
a part or particle considered to be an irreducible constituent of a specified system	
having a volume equal to a cube whose edge is of a stated length	
an electric generator that produces alternating current	
of or relating to magnetism or magnets	
the force which causes things to drop to the ground	
a unit of potential difference equal to one thousandth of a volt	
a building or group of buildings for the manufacture of a product	

【Ex.3】把下列句子翻译成中文。

1. A glass rod becomes charged when rubbed with silk, as does a hard-rubber rod when rubbed with fur.

2. People find that charges produce forces of repulsion and attraction.

3. Negative charges repel negative charges, positive charge repel positive charges, and positive and negative charges attract each other.

4. Scientists find that all negative charges are integer multiples of a certain very small charge.

5. Positive charges are integer multiples of a very small charge.

6. In general, charge cannot be created or destroyed, a fact called the Law of Conservation of Charge.

7. Rubbing the glass rod with a silk cloth removes electrons from the rod and puts them on the cloth.

8. Electrons of an atom have orbits at different distances from the nucleus.

9. Ions are charged particles that would produce a current if they could move.

10. Even at normal room temperatures the outer electrons in metals receive enough heat energy to become free, especially for silver, copper, gold, aluminum.

【Ex.4】把下列短文翻译成中文。

The charge of an electron or of a proton is much too small to be basic quantity of charge for almost all practical applications. The SI unit of charge is the coulomb, with the symbol C. A coulomb of negative charge equals that of 6.242×10^{18} electrons. The coulomb is a derived SI unit, which means that it can be derived from SI base units.

【Ex.5】通过 Internet 查找资料，借助电子词典、辅助翻译软件及 AI 工具，完成以下技术报告，并附上收集资料的网址。通过 E-mail 发送给老师，或按照教学要求在网上课堂提交。
1. 当前世界上有哪些流行的工业控制软件，以及它们的主要技术指标（附各种最新产品的主要界面图片）。
2. 当前世界上有哪些流行的电路设计软件，以及它们的主要技术指标（附各种最新产品的主要界面图片）。

Text B

Introduction to AC

The supply of current for electrical devices may come from a direct current source (DC), or an alternating current source (AC). In direct current electricity, electrons flow continuously in one direction from the source of power through a conductor to a load and back to the source of power. The voltage in direct current remains constant. DC power sources include batteries and DC generators. In alternating current an AC generator is used to make electrons flow first in one direction then in another. Another name for an AC generator is an alternator. The AC generator reverses terminal polarity many times a second. Electrons will flow through a conductor from the negative terminal to the positive terminal, first in one direction then another (see Figure 1-9).

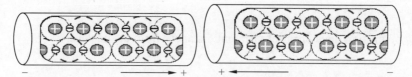

Figure 1-9 Electrons will flow through a conductor from the negative terminal to the positive terminal

1. AC

1.1 AC sine wave

Alternating voltage and current vary continuously. The graphic representation for AC is a sine wave. A sine wave can represent current or voltage. There are two axes. The vertical axis represents the direction and magnitude of current or voltage. The horizontal axis represents time (see Figure 1-10).

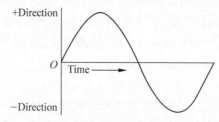

Figure 1-10 The graphic representation for AC is a sine wave

When the waveform is above the time axis, current is flowing in one direction. This is referred to as the positive direction. When the waveform is below the time axis, current is flowing in the opposite direction. This is referred to as the negative direction. A sine wave moves through a complete rotation of 360°, which is referred to as one cycle. Alternating current goes through many of these cycles each second. The unit of measurement of cycles per second is hertz(Hz). In the United States alternating current is usually generated at 60Hz.

1.2 Single-phase and three-phase AC power

Alternating current is divided into single-phase and three-phase types. Single-phase power is used for small electrical demands such as found in the home. Three-phase power is used where large blocks of power are required, such as found in commercial applications and industrial plants. Three-phase power, as shown in the following illustration, is a continuous series of three overlapping AC cycles. Each wave represents a phase, and is offset by 120 electrical degrees (see Figure 1-11).

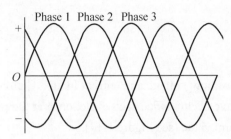

Figure 1-11　Three-phase AC power

1.3　Frequency

The number of cycles per second made by voltage induced in the armature is decided by the frequency of the generator. If the armature rotates at a speed of 60 revolutions per second, the generated voltage will be 60 cycles per second. The accepted term for cycles per second is hertz. The standard frequency in the United States is 60Hz. The following illustration shows 15 cycles in 1/4 second which is equivalent to 60 cycles in one second (see Figure 1-12).

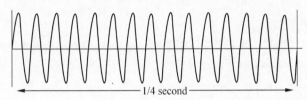

Figure 1-12　Frequency

1.4　Four-pole AC generator

The frequency is the same as the number of rotations per second if the magnetic field is produced by only two poles. An increase in the number of poles will cause an increase in the number of cycles completed in a revolution. A two-pole generator will complete one cycle per revolution and a four-pole generator will complete two cycles per revolution. An AC generator produces one cycle per revolution for each pair of poles (see Figure 1-13).

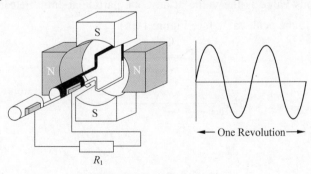

Figure 1-13　AC generator

2. Voltage and Current

2.1　Peak value

The sine wave illustrates how voltage and current in an AC circuit rises and falls with time. The peak value of a sine wave occurs twice each cycle, once at the positive maximum value and once at the negative maximum value (see Figure 1-14).

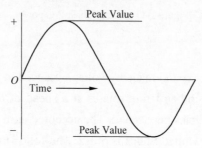

Figure 1-14　Peak value

2.2　Peak-to-peak value

The value of the voltage or current between the peak positive and peak negative values is called the peak-to-peak value (see Figure 1-15).

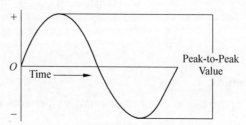

Figure 1-15　Peak-to-peak value

2.3　Instantaneous value

The instantaneous value is the value at any one particular time. It can be in the range of anywhere from zero to the peak value (see Figure 1-16).

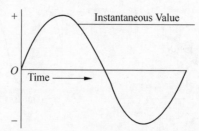

Figure 1-16　Instantaneous value

2.4 Calculating instantaneous voltage

The voltage waveform produced as the armature rotates through 360° rotation is called a sine wave because instantaneous voltage is related to the trigonometric function called sine ($\sin\theta$ = sine of the angle). The sine curve represents a graph of the following equation:

$$e = E_{peak} \times \sin\theta$$

Instantaneous voltage is equal to the peak voltage times the sine of the angle of the generator armature. The sine value is obtained from trigonometric tables. The following Table 1-3 reflects a few angles and their sine values.

Table 1-3 A few angles and their sine values

Angle(θ)/(°)	$\sin\theta$	Angle(θ)/(°)	$\sin\theta$
30	0.5	210	−0.5
60	0.866	240	−0.866
90	1	270	−1
120	0.866	300	−0.866
150	0.5	330	−0.5
180	0	360	0

The following example illustrates instantaneous values at 90°, 150° and 240° (see Figure 1-17). The peak voltage is equal to 100V. By substituting the sine at the instantaneous angle value, the instantaneous voltage can be calculated.

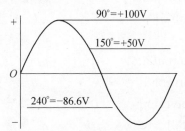

Figure 1-17 Example of instantaneous values

Any instantaneous value can be calculated. For example 240°

$$e = 100 \times (-0.866)$$
$$= -86.6V$$

2.5 Effective value of an AC sine wave

Alternating voltage and current are constantly changing values. A method of translating the varying values into an equivalent constant value is needed. The effective value of voltage and current is the common method of expressing the value of AC. This is also known as the RMS (root-mean-square) value. If the voltage in the average home is said to be 120V, this is the RMS value. The effective value figures out to be 0.707 times the peak value (see Figure 1-18).

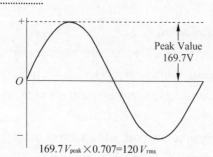

Figure 1-18　Effective value of an AC sine wave

　　The effective value of AC is defined in terms of an equivalent heating effect when compared to DC. One RMS ampere of current flowing through a resistance will produce heat at the same rate as a DC ampere. For purpose of circuit design, the peak value may also be needed. For example, insulation must be designed to withstand the peak value, not just the effective value. It may be that only the effective value is known. To calculate the peak value, multiply the effective value by 1.41. For example, if the effective value is 100V, the peak value is 141V.

New Words

reverse	[rɪˈvɜːs]	n. 逆向；逆转
polarity	[pəˈlærəti]	n. 极性
sine	[saɪn]	n. 正弦
continuously	[kənˈtɪnjʊəslɪ]	adv. 不断地，连续地
vertical	[ˈvɜːtɪkl]	adj. 垂直的 n. 垂直线，垂直面，竖向
magnitude	[ˈmæɡnɪtjuːd]	n. 大小，数量
horizontal	[ˌhɒrɪˈzɒntl]	adj. 水平的
waveform	[ˈweɪvfɔːm]	n. 波形
cycle	[ˈsaɪkl]	n. 周期
hertz	[hɜːts]	n. 赫（Hz），赫兹
phase	[feɪz]	n. 相，相位
three-phase	[θriːˈfeɪz]	adj. 三相的
offset	[ˈɒfset]	n. 偏移量
frequency	[ˈfriːkwənsɪ]	n. 频率
overlap	[əʊvəˈlæp]	vt. & vi. 重叠绕包，搭接
armature	[ˈɑːmətʃə]	n. 电枢（电机的部件）
pole	[pəʊl]	n. 极，磁极，电极
peak	[piːk]	n. 顶点
maximum	[ˈmæksɪməm]	n. 最大量，最大值，极大，极大值
instantaneous	[ˌɪnstənˈteɪnɪəs]	adj. 瞬间的，即刻的，即时的
trigonometric	[ˌtrɪɡənəˈmetrɪk]	adj. 三角法的，据三角法的

effective	[ɪˈfektɪv]	*adj.* 有效的
root-mean-square	[ruːt miːn skweə]	*n.* 均方根（值）
withstand	[wɪðˈstænd]	*vt.* 经受住，抵挡

Phrases

graphic representation	图示
sine wave	正弦波
be referred to as	称为
positive direction	正向
opposite direction	反向
negative direction	负向
magnetic field	磁场
divide into	分为
peak value	峰值
trigonometric function	三角函数

Abbreviations

AC (Alternating Current)	交流电
DC (Direct Current)	直流电

Exercises

【**Ex.6**】根据文章所提供的信息判断正误。

1. In direct current electricity, electrons flow continuously in one direction from the source of power through a conductor to a load and back to the source of power.
2. In alternating current an AC generator is used to make electrons flow continuously in one direction.
3. Electrons will flow through a conductor from the positive terminal to the negative terminal.
4. The graphic representation for AC is a sine wave.
5. In China alternating current is usually generated at 60Hz.
6. A sine wave moves through a complete rotation of 360°, which is referred to as one cycle.
7. The peak value of a sine wave occurs once each cycle.
8. The instantaneous value is the value at any one particular time. It can be in the range of anywhere from zero to the peak value.
9. Instantaneous voltage is equal to the peak voltage times the angle of the generator armature.
10. For example, if the effective value is 110V, the peak value is 220V.

科技英语翻译知识

翻译的标准

翻译标准（translation norm, translation criteria, translation standard principle）是翻译实践的准绳和衡量译文好坏的尺度，也是翻译工作者要努力达到的目标。中外翻译理论家历来认为这是翻译理论的核心问题，因此他们提出了种种论述。

在我国，早在三国时期，支谦就提出了"循本旨，不加文饰"的译经原则。唐代的玄奘提出了"既须求真，又须喻俗"。不过在翻译界影响最大、最强烈的还是清末著名翻译家严复提出的"信、达、雅"三字原则或三字标准。当代的翻译家又有多种说法，有鲁迅的"信、和、顺"标准、朱生豪的"神韵说"、傅雷的"神似说"、钱钟书的"化境说"、许渊冲的"三美说"等。

国外也有多种论述，18世纪的英国学者、爱丁堡大学教授泰特勒（A. F. Tytler）提出翻译三原则：要完整再现原文；保留原文的风格和手法；译文应和原文一样通顺。

苏联翻译家费道罗夫提出了"等值论"。当代美国翻译家尤金·A. 奈达（Eugene A. Nida）提出了"等效论"，又称动态对等说（dynamic equivalence）或功能对等论（functional equivalence）。

国内外的专家对翻译标准有这么多的论述，令初学者感到眼花缭乱，正因为如此，有的专家提出简单明了、普遍能为人们接受的翻译标准，即"忠实通顺"。

科技英语虽有其自身的特点，但是在翻译标准上并不能例外。"忠实通顺"也应当作为科技英语的翻译标准。

1. 忠实

忠实指忠于原作的内容，要完整而准确地表达出来，不仅要忠于原作的思想、观点、立场和所流露的感情，还要忠于原作的风格，即原作的民族风格、时代风格、作者个人的语言风格等。总之，原作怎样，译文也应该怎样，尽可能还其本来面目。正如鲁迅所说的，"保存原作的风姿"。

下面通过例句来说明对忠实的理解：

The engine did not stop because the fuel was finished.

译文是"发动机没有停止，因为燃料用完了"。这样的译文虽然"完全"忠于原文，但只是忠于原作的形式，并没有忠于原作的内容。译文显然不符合逻辑——这是因为译者缺乏对原文的理解：这是英语中的否定转移，not 并不是修饰谓语动词 stop，而是修饰句子的后面部分。正确的译文应该是"发动机不是因为燃料用完而停止的"。

又如：

All substances will permit the passage of some electric current, provided the potential difference is high enough.

句中 all substances 相对于 permit 来说是事，但是对于后面的 passage 来说却是地点。因为人们可以将 some electric current、passage 和 all substances 理解为 some electric current pass（through）all substances。这是因为在科技英语中的复合名词词组中，其深层结构中的语义关系比较复杂。如果忽略了这一点，只按照表层结构理解，译文是"只要有足够的电位差，所有的物体都允许一些电流通过"。如果从语义关系来考虑，译文应是"只要有足够的电位差，电流便可以通过任何物体"。

可见忠实并不仅仅是忠于文字的表面，而是忠于语言的内容。

2．通顺

通顺指译文必须通顺易懂，符合规范。具体地说，科技英语的翻译要符合科技语言的规范，要用明白晓畅的现代语言，没有逐词死译、硬译的现象，没有语言晦涩、文理不通、结构混乱、逻辑不清的现象。下面通过实例来说明对通顺的理解：

The virtual reality technology is hindered right now by the fact that today's computers are simply not fast enough.

译文是"虚拟现实技术被今天计算机不够快速所制约"。译者试图尽量忠于原文，但是这样的译文读起来非常别扭，原因是没有注意到英汉语言在表达上的差异，所以一点也不通顺。正确的译文应该是"目前，计算机运行速度缓慢制约着虚拟现实技术的发展"。这个译文对原文的次序做了调整，还增加了原文中形无而实有的"运行速度"一词；把原文的被动语态改为汉语中的主动语态。此译文不仅忠于原文的内容，也符合汉语的规范。

忠实和通顺是相辅相成的。忠实而不通顺，读者看不懂，也就谈不到忠实；通顺而不忠实，也就使译文失去了原意，成为杜撰。

所以，在科技英语的翻译中，以"忠实通顺"作为翻译标准是切实可行的。

Reading Material

阅读下列文章。

Text	Note
Introduction to Electrical Engineering	
1. Introduction	
Electrical engineering is an engineering discipline that uses electrical principles to design, manufacture and maintain various electrical equipment and systems. These devices and systems include various fields such as power transmission and distribution systems, power generation, electromechanical[1] control, automation control, communication systems, computer networks, and electronic equipment. Electrical engineering mainly studies the transmission, conversion and control of electrical energy, which has a wide range	[1] *adj.* 电动机械的，机电的，电机的

of applications in modern society, including transportation[2], communication, medical treatment, construction, manufacturing and other fields.	[2] *n.* 运输
Electrical engineering includes multiple sub-disciplines, such as circuit theory, electromagnetic field theory, power system, motor and transformer[3], electronic technology, control theory, etc. These sub-disciplines are interrelated and form the core body of knowledge in electrical engineering. The application fields of electrical engineering are also very wide, including power engineering, communication engineering, computer engineering, automation engineering, manufacturing engineering, etc.	[3] *n.* 变压器
The development of electrical engineering has promoted the progress of modern industry and technology, such as the invention and application of electric motors, the construction and management of power systems, the development of communication technology, the application of computers, and so on. With the continuous advancement of science and technology, the development of the field of electrical engineering is also expanding and deepening, and the application scope and development prospects[4] of electrical engineering will continue to expand in the future.	[4] *n.* 前景，期望

2. Research Direction of Electrical Engineering

2.1 Power System and Its Automation

Power system and its automation is one of the most important research directions in the field of electrical engineering. In this field, we can delve into the design, operation, maintenance and other aspects of power systems. In addition, we can also study how automation technology can be used to improve the efficiency and stability[5] of power systems. During the study, the power system can be analyzed using modeling and simulation techniques for optimal control and monitoring.	[5] *n.* 稳定性

2.2 Power Electronics and Electric Drive Systems

Power electronics technology is another important research direction in the field of electrical engineering. In this field, we can study the design, working principle and other aspects of power electronic devices. At the same time, we can also study the application of power electronics in motor drive and generator[6] control to improve the efficiency and reliability of the electric drive systems.

2.3 Smart Grid and New Energy Systems

Smart grid and new energy systems are emerging research directions in the field of electrical engineering in recent years. In this field, we can study the design, operation, control and other aspects of smart grid. At the same time, we can also conduct in-depth research on the application and development of new energy systems, such as solar energy, wind energy, water energy, etc., to replace traditional energy sources.

2.4 Communication and Signal Processing Technology

Communication and signal processing technology is one of the important research directions in the field of electrical engineering. In this field, we can study communication principles, signal processing techniques, communication networks[7], etc. At the same time, we can also deeply study the application of artificial intelligence, machine learning and other technologies in the field of signal processing to improve the efficiency and quality of communication systems.

2.5 Microelectronics Technology and Integrated Circuit Design

Microelectronics[8] technology and integrated circuit design are another important research direction in the field of electrical engineering. In this field, we can study semiconductor devices, microelectronics manufacturing, integrated circuit design, packaging technology, etc. In the research process, we can use EDA software to design and simulate the circuit for optimization and improvement.

3. Employment Orientation of Electrical Engineering Majors

(1) Automation control engineer: engaged in automation system design, controller programming, process optimization, etc., applied to manufacturing, energy industry, transportation and other fields.

(2) Power system engineer: Responsible for power system planning, design, operation and maintenance, etc., and participating[9] in power grid construction and energy scheduling.

(3) Electronic engineer: engaged in electronic product design, research and development, testing, etc., involving electronic circuits,

---

[6] *n.* 发电机,发生器

[7] *n.* 网络

[8] *n.* 微电子学

[9] *v.* 参与,参加

communication systems, embedded systems, etc.

(4) Communication engineer: Responsible for the design, construction and maintenance of communication networks, including wireless[10] communication, optical fiber communication, etc.

(5) Control system engineer: Participating in the design, integration and optimization of industrial automation control systems, which are applied in manufacturing, process control and other fields.

(6) Intelligent manufacturing and Internet of Things engineers: engaged in the research and development and application of intelligent manufacturing equipment and Internet of Things systems, and promoting industrial intelligence and digital transformation.

(7) New energy engineers: participating in the research and development, application and management of new energy technologies, and promoting the utilization[11] and development of renewable energy.

(8) Environmental monitoring and treatment[12] engineer: Responsible for the research and development and application of environmental monitoring equipment and treatment technology, and promoting the development of the environmental protection industry.

4. The Development Direction of the Electrical Engineering Industry

First of all, with the increasingly severe global energy problems, the field of new energy will become an important development direction of the electrical engineering industry. With the continuous development of solar energy, wind energy and other new energy technologies, it will bring a lot of employment opportunities and business opportunities.

Secondly, artificial intelligence technology will also become an important development direction of the electrical engineering industry. With the continuous development of artificial intelligence technology, there will be more automation and intelligent equipment, which need to be based on the application of electrical engineering technology, so it will also bring more opportunities and challenges[13].

In addition, the electrical engineering industry will continue to benefit from the development of Internet technology. With the continuous development of technologies such as the Internet of Things and cloud computing, there will be more digital and intelligent

[10] *adj.* 无线的

[11] *n.* 利用,使用

[12] *n.* 处理

[13] *n.* 挑战

applications in the electrical engineering industry, which will bring more business opportunities and employment opportunities.

参 考 译 文

电的基本概念

1. 电荷

1.1 原子的中性态

元素通常通过组成该元素的原子核周围轨道上的电子数量和原子核中的质子数量来识别。例如，一个氢原子只有一个电子和一个质子。一个铝原子有 13 个电子和 13 个质子，如图 1-1 所示。一个有相同数目的电子和质子数的原子电性呈中性。

（图略）

1.2 正负电荷

原子外圈的电子很容易由于受到外力的作用而被移走。电子从轨道移走会造成该原子缺少电子，而移入电子的原子会使电子过剩。由于质子数多于电子数，因此缺少电子的原子称为正电荷。电子过剩的原子带有负电荷。电子不足或过剩产生正电荷或负电荷。质子数始终是恒定不变的，如图 1-2 所示。

（图略）

1.3 电荷之间的吸引和排斥

古语"异性相吸"在处理电荷时是正确的。每个电荷四周都有一个看不见的电场。当两个电性相同的电荷靠近时，电场将使两电荷相斥，当两个电性不同的电荷靠近时，电场将使两电荷相吸。电荷周围的电场用不可见的电力线表示，不可见的电力线表示引起吸引和排斥的不可见的电场。正电荷用离开的电力线表示，负电荷用进入的电力线表示，如图 1-3 所示。

（图略）

2. 电流

电是在导体中的从一个原子以相同方向流向下一个原子的自由电子流。这种电子流就是电流，由符号"I"表示。电子以不同的速度流过导体，电流有不同的值。电流的大小由在 1s 内流过导体横截面的电子数量决定。我们必须记住原子是非常小的。在 $1cm^3$ 的铜导体中约有 1 000 000 000 000 000 000 000 000 个原子。这个数字可以用数学的指数形式简化，不用写 1 后的 24 个 0，而写成 10^{24}。试图测量即使是很小的电流也将产生难以想象的巨大的数字。所以，电流用安培数来度量，安培的英文可以简写为"amp"。安培可以用符号"A"表示。1A 的电流意味着在 1s 之内有 6.24×10^{18} 个电子流过导体的截面。这些数字只是信息，并不需要去关心。但是重要的是理解电流的概念，如图 1-4 所示。

（图略）

2.1 测量电流的单位

表 1-1 反映了处理很小和很大的电流时计量单位的词头：

（表略）

2.2 电流的方向

一些权威将电子流和电流区分，传统上电流理论忽略电子流，而认为电流是从正极流向负极。为了避免混淆，我们使用的电子流概念是电子从负极流向正极，如图 1-5 所示。

（图略）

3. 电压

可以把电比作流过管子的水，如图 1-6 所示。水流过管子需要一个力的作用，这个力来自水泵或水的重力。电压就是作用于导体上的引起电流流动的力。电子的电性是负的并能被正电荷吸引，电子总是从由于多余电子而呈负电荷的地方被吸引到由于缺乏电子而呈现电荷为正的地方。这种使电子流过导体的力叫作电势差、电动势或简单地称为电压。电压用字符"U"或"V"表示。度量电压的单位是伏特，伏特用字符"V"表示。

（图略）

3.1 电压源

可以有多种产生电压的方式。电池利用电化学过程，而汽车的发电机和电厂的发电机利用电磁感应作用。所有的电压源共有的特性是在一端电子过剩并在另一端电子不足，这就导致了在两端具有不同的电动势，如图 1-7 所示。

（图略）

3.2 电压符号

在电路图中画两条线来象征性代表电池的两端。长的线代表电池的正极，短的线代表电池的负极（见图 1-8）。

（图略）

3.3 电压度量单位

表 1-2 反映了处理很小和很大的电压时计量单位的词头：

（表略）

Text A

Electrical Resistance and Conductance

The electrical resistance of an electrical conductor is a measure of the difficulty to pass an electric current through that conductor. The inverse quantity is electrical conductance, and is the ease with which an electric current passes. Electrical resistance shares some conceptual parallels with the notion of mechanical friction. The IS unit of electrical resistance is the ohm (Ω), while electrical conductance is measured in siemens (S).

An object of uniform cross section has a resistance proportional to its resistivity and length and inversely proportional to its cross-sectional area. All materials show some resistance, except for superconductors, which have a resistance of zero[1].

The resistance of an object is defined as the ratio of voltage across it (V) to current through it (I), while the conductance (G) is the inverse:

$$R = V / I \qquad G = I / V = 1 / R$$

For a wide variety of materials and conditions, V and I are directly proportional to each other, and therefore R and G are constant (although they can depend on other factors like temperature). This proportionality is called Ohm's law, and materials that satisfy it are called ohmic materials.

In other cases, such as a diode or battery, V and I are not directly proportional. The ratio V/I is sometimes still useful, and is referred to as a "static resistance", since it corresponds to the inverse slope of a chord between the origin and an I–V curve. In other situations, the derivative dV/dI may be most useful; this is called the "differential resistance".

1. Introduction

In the hydraulic analogy, current flowing through a wire (or resistor) is like water flowing through a pipe, and the voltage drop across the wire is like the pressure drop that pushes water through the pipe[2] (see Figure 2-1). Conductance is proportional to how much flow occurs for a given pressure, and resistance is proportional to how much pressure is required to achieve a given flow. Conductance and resistance are reciprocals.

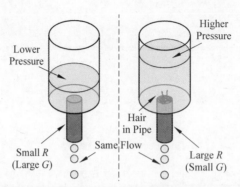

Figure 2-1　The hydraulic analogy compares electric current flowing through circuits to water flowing through pipes. When a pipe (left) is filled with hair (right), it takes a larger pressure to achieve the same flow of water. Pushing electric current through a large resistance is like pushing water through a pipe clogged with hair: It requires a larger push (electromotive force) to drive the same flow (electric current).

　　The voltage drop (i.e., difference between voltages on one side of the resistor and the other), not the voltage itself, provides the driving force pushing current through a resistor. In hydraulics, it is similar: The pressure difference between two sides of a pipe, not the pressure itself, determines the flow through it. For example, there may be a large water pressure above the pipe, which tries to push water down through the pipe. But there may be an equally large water pressure below the pipe, which tries to push water back up through the pipe. If these pressures are equal, no water flows.

　　The resistance and conductance of a wire, resistor, or other element is mostly determined by two properties:

　　(1) geometry (shape), and

　　(2) material.

　　Geometry is important because it is more difficult to push water through a long, narrow pipe than a wide, short pipe. In the same way, a long, thin copper wire has higher resistance (lower conductance) than a short, thick copper wire.

　　Materials are important as well. A pipe filled with hair restricts the flow of water more than a clean pipe of the same shape and size. Similarly, electrons can flow freely and easily through a copper wire, but cannot flow as easily through a steel wire of the same shape and size, and they essentially cannot flow at all through an insulator like rubber, regardless of its shape. The difference between, copper, steel, and rubber is related to their microscopic structure and electron configuration, and is quantified by a property called resistivity.

2. Conductors and Resistors

　　Substances in which electricity can flow are called conductors. A piece of conducting material of a particular resistance meant for use in a circuit is called a resistor (see Figure 2-2). Conductors are made of high-conductivity materials such as metals, in particular copper and aluminium. Resistors, on the other hand, are made of a wide variety of materials depending on factors such as the desired resistance, amount of energy that it needs to dissipate, precision, and costs[3].

Figure 2-2 A 6.5 MΩ resistor, as identified by its electronic color code (blue–green–black-yellow). An ohmmeter could be used to verify this value.

3. Relation to Resistivity and Conductivity

The resistance of a given object depends primarily on two factors: What material it is made of, and its shape. For a given material, the resistance is inversely proportional to the cross-sectional area; for example, a thick copper wire has lower resistance than an otherwise- identical thin copper wire. Also, for a given material, the resistance is proportional to the length; for example, a long copper wire has higher resistance than an otherwise-identical short copper wire. The resistance R and conductance G of a conductor of uniform cross section, therefore, can be computed as

$$R=\rho(L/A)$$
$$G=\sigma(A/L)$$

where L is the length of the conductor, measured in metres[m], A is the cross-sectional area of the conductor measured in square metres[m²], σ (sigma) is the electrical conductivity measured in siemens per meter (S·m^{-1}), and ρ is the electrical resistivity (also called specific electrical resistance) of the material, measured in ohm-metres (Ω·m). The resistivity and conductivity are proportionality constants, and therefore depend only on the material the wire is made of, not the geometry of the wire. Resistivity and conductivity are reciprocals: $\rho= 1 /\sigma$. Resistivity is a measure of the material's ability to oppose electric current.

This formula is not exact, as it assumes the current density is totally uniform in the conductor, which is not always true in practical situations. However, this formula still provides a good approximation for long thin conductors such as wires.

Another situation for which this formula is not exact is with alternating current (AC), because the skin effect inhibits current flow near the center of the conductor. For this reason, the geometrical cross-section is different from the effective cross-section in which current actually flows, so resistance is higher than expected. Similarly, if two conductors near each other carry AC current, their resistances increase due to the proximity effect. At commercial power frequency, these effects are significant for large conductors carrying large currents, such as busbars in an electrical substation, or large power cables carrying more than a few hundred amperes[4].

4. Measuring Resistance

An instrument for measuring resistance is called an ohmmeter. Simple ohmmeters cannot measure low resistances accurately because the resistance of their measuring leads causes a voltage drop that interferes with the measurement, so more accurate devices use four-terminal sensing.

New Words

conductance	[kənˈdʌktəns]	n.	电导，导体，电导系数
electrical	[ɪˈlektrɪkl]	adj.	电的，有关电的
conductor	[kənˈdʌktə]	n.	导体
ohm	[əʊm]	n.	欧姆
siemens	[ˈsiːmənz]	n.	西门子（欧姆的倒数）
resistivity	[ˌrɪzɪˈstɪvɪtɪ]	n.	电阻系数
proportional	[prəˈpɔːʃənl]	adj.	比例的，成比例的，相称的
superconductor	[ˈsuːpəkəndʌktə]	n.	超导（电）体
voltage	[ˈvəʊltɪdʒ]	n.	电压，伏特数
constant	[ˈkɒnstənt]	n.	常数，恒量
		adj.	不变的
temperature	[ˈtemprətʃə]	n.	温度
proportionality	[prəˌpɔːʃəˈnælətɪ]	n.	比例（性）
ohmic	[ˈəʊmɪk]	adj.	欧姆的
diode	[ˈdaɪəʊd]	n.	二极管
slope	[sləʊp]	n.	斜坡，斜面，倾斜，斜度，斜率
chord	[kɔːd]	n.	弦
curve	[kɜːv]	n.	曲线，弯曲
		vt.	弯，使弯曲
		vi.	成曲形
analogy	[əˈnælədʒɪ]	n.	模拟，类推
reciprocal	[rɪˈsɪprəkl]	adj.	倒数的
		n.	倒数
resistor	[rɪˈzɪstə]	n.	电阻器
hydraulics	[haɪˈdrɔːlɪks]	n.	水力学
geometry	[dʒɪˈɒmətrɪ]	n.	几何形状
shape	[ʃeɪp]	n.	外形，形状
narrow	[ˈnærəʊ]	n.	狭窄部分，隘路
		adj.	狭窄的
thick	[θɪk]	adj.	厚的，粗的
electron	[ɪˈlektrɒn]	n.	电子
insulator	[ˈɪnsjuleɪtə]	n.	绝缘体，绝热器
rubber	[ˈrʌbə]	n.	橡皮，橡胶
microscopic	[ˌmaɪkrəˈskɒpɪk]	adj.	用显微镜可见的，精微的
ohmmeter	[ˈəʊmmiːtə]	n.	欧姆计，电阻表

verify	[ˈverɪfaɪ]	vt. 检验，校验，查证，核实
substance	[ˈsʌbstəns]	n. 物质
conductivity	[ˌkɒndʌkˈtɪvətɪ]	n. 传导性，传导率
aluminium	[ˌæljəˈmɪnɪəm]	n. 铝 adj. 铝的
dissipate	[ˈdɪsɪpeɪt]	v. 消耗
precision	[prɪˈsɪʒn]	n. 精确，精密度，精度
thin	[θɪn]	adj. 薄的，细的
identical	[aɪˈdentɪkl]	adj. 同一的，同样的
compute	[kəmˈpjuːt]	v. 计算，估计
sigma	[ˈsɪgmə]	n. 西格玛，希腊字母(Σ, σ)
oppose	[əˈpəʊz]	v. 抵制，反对
exact	[ɪgˈzækt]	adj. 精确的，准确的
formula	[ˈfɔːmjələ]	n. 公式，规则
inhibit	[ɪnˈhɪbɪt]	v. 抑制，约束
busbar	[ˈbʌzbɑː]	n. 母线，母排；汇流排，汇流条
ampere	[ˈæmpeə]	n. 安培
instrument	[ˈɪnstrəmənt]	n. 工具，器械，器具

Phrases

electric current	电流
uniform cross section	等截面
cross-sectional area	断面面积，横截面积
be defined as	被定义为
a variety of	多种的
Ohm's law	欧姆定律
static resistance	静态电阻
differential resistance	微分电阻，内阻
electromotive force	电动势
pressure drop	压力降
driving force	驱动力
pressure difference	压力差，差压
copper wire	铜线
fill with	使充满
steel wire	钢丝
regardless of	不管，不顾
electron configuration	电子构型，电子组态
be made of	用……制造，形成，构成
square metres	平方米

electrical conductivity	电导率
proportionality constant	比例常数
current density	电流密度
practical situation	实际情况
skin effect	趋肤效应
proximity effect	邻近效应
commercial power frequency	市电频率
electrical substation	变压站
power cable	电力电缆
interfere with	妨碍，干涉，干扰
four-terminal sensing	四端子检测，四线检测，四点探针法

Abbreviations

IS (International System)　　　　　　国际系统

Notes

[1] All materials show some resistance, except for superconductors, which have a resistance of zero.

本句中，which have a resistance of zero 是一个非限定性定语从句，对 superconductors 进行补充说明。except for 的意思是"除……之外"。

本句意为：除具有零电阻的超导体之外，所有材料都有一定量的电阻。

[2] In the hydraulic analogy, current flowing through a wire (or resistor) is like water flowing through a pipe, and the voltage drop across the wire is like the pressure drop that pushes water through the pipe.

本句中，flowing through a wire (or resistor)是一个现在分词短语，作定语，修饰和限定 current；flowing through a pipe 也是一个现在分词短语，作定语，修饰和限定 water。that pushes water through the pipe 是一个定语从句，修饰和限定 the pressure drop。

本句意为：在液压模拟中，流过导线（或电阻器）的电流类似于流经管道的水，穿过导线的电压降类似于推动水通过管道的压力降。

[3] Resistors, on the other hand, are made of a wide variety of materials depending on factors such as the desired resistance, amount of energy that it needs to dissipate, precision, and costs.

本句中，on the other hand 的意思是"另一方面"；be made of 的意思是"由……构成"；a wide variety of 的意思是"种种，多种多样"。that it needs to dissipate 是一个定语从句，修饰和限定 amount of energy。such as the desired resistance, amount of energy that it needs to dissipate, precision, and costs 对 factors 进行举例说明。

本句意为：另一方面，电阻器由多种材料制成，这取决于期望的电阻、需要消耗的能量、精度和成本等因素。

[4] At commercial power frequency, these effects are significant for large conductors carrying

large currents, such as busbars in an electrical substation, or large power cables carrying more than a few hundred amperes.

本句中，carrying large currents 是一个现在分词短语，作定语，修饰和限定 large conductors, such as busbars in an electrical substation 是对 large conductors 的举例说明；carrying more than a few hundred amperes 也是一个现在分词短语，作定语，修饰和限定 large power cables。

本句意为：在商业电源频率下，这些效应对于承载大电流的大导体（如变电站中的母线或承载数百安培以上的大功率电缆）是显著的。

Exercises

【Ex.1】根据课文内容，回答以下问题。

1. What is the electrical resistance of an electrical conductor?

2. What provides the driving force pushing current through a resistor? What does the pressure difference between two sides of a pipe determine?

3. What determines the resistance and conductance of a wire, resistor, or other element?

4. What are conductors made of?

5. Why cannot simple ohmmeters measure low resistances accurately?

【Ex.2】根据下面的英文解释，写出相应的英文词汇。

英 文 解 释	词 汇
the degree to which a substance prevents the flow of an electric current through it	
a material or covering which electricity, heat or sound cannot go through	
a substance that allows electricity or heat to pass along it or through it	
the standard unit of electrical resistance	
the standard unit used to measure how strongly an electrical current is sent around an electrical system	
the state that two pieces of wire are connected end to end	
the state that two pieces of wire were connected side by side	
the standard unit of measurement for the strength of an electrical current	

英 文 解 释	词 汇
a standard or accepted way of doing or making something, the items needed for it, or a mathematical rule expressed in a set of numbers and letters	
the standard measure of electrical power	

【Ex.3】把下列句子翻译成中文。

1. Resistors are used to control voltages and currents.

2. Resistance determines how much current will flow through a component.

3. A very high resistance allows very little current to flow.

4. Sparks and lightning are brief displays of current flow through air.

5. A low resistance allows a large amount of current to flow.

6. Wires are usually covered with rubber or plastic.

7. High voltage power lines are covered with thick layers of plastic to make them safe.

8. Making the resistance higher will let less current flow.

9. Variable resistors have a dial or a knob that allows you to change the resistance.

10. A 500Ω variable resistor can have a resistance of anywhere between 0Ω and 500Ω.

【Ex.4】把下列短文翻译成中文。

Resistance is given in units of ohm(Ω). (Ohm is named after Georg Simon Ohm who played with electricity as a young boy in Germany.) Common resistor values are from 100Ω to 100 000Ω.

Each resistor is marked with colored stripes to indicate its resistance. You can determine the resistance value of the resistor by looking at the stripes on it.

【Ex.5】通过 Internet 查找资料，借助电子词典、辅助翻译软件及 AI 工具，完成以下技术报告，并附上收集资料的网址。通过 E-mail 发送给老师，或按照教学要求在网上课堂提交。
1. 当前世界上有哪些最主要的电阻生产厂家及有哪些最新型的产品（附各种最新产品的图片）。
2. 当前世界上有哪些最主要的电容生产厂家及有哪些最新型的产品（附各种最新产品的图片）。

Text B

Capacitor

A capacitor is a little like a battery but works completely differently. A battery is an electronic device that converts chemical energy into electrical energy, whereas a capacitor is an electronic component that stores electrostatic energy in an electric field.

1. What Is a Capacitor?

A capacitor is a two-terminal electrical device that can store energy in the form of an electric charge. It consists of two electrical conductors that are separated by a distance. The space between the conductors may be filled by vacuum or with an insulating material known as a dielectric. The ability of the capacitor to store charges is known as capacitance.

Capacitors store energy by holding apart pairs of opposite charges. The simplest design for a capacitor is a parallel plate, which consists of two metal plates with a gap between them. But, different types of capacitors are manufactured in many forms, styles, lengths, girths, and materials.

2. How Does a Capacitor Work?

For demonstration, let us consider the most basic structure of a capacitor — the parallel plate capacitor (see Figure 2-3). It consists of two parallel plates separated by a dielectric. When we connect a DC voltage source across the capacitor, one plate is connected to the positive end (plate I) and the other to the negative end (plate II). When the potential of the battery is applied across the capacitor, plate I become positive with respect to plate II. The current tries to flow through the capacitor at the steady-state condition from its positive plate to its negative plate. But it cannot flow due to the separation of the plates with an insulating material.

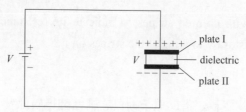

Figure 2-3　A battery is connected across a parallel plate capacitor

An electric field appears across the capacitor (see Figure 2-4). The positive plate (plate I) accumulates positive charges from the battery, and the negative plate (plate II) accumulates negative charges from the battery. After a point, the capacitor holds the maximum amount of charge as per its capacitance with respect to this voltage. This time span is called the charging time of the capacitor.

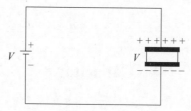

Figure 2-4　An electric field appears across the capacitor

When the battery is removed, the two plates hold a negative and positive charge for a certain time. Thus, the capacitor acts as a source of electrical energy.

If these plates are connected to a load, the current flows to the load from plate I to plate II until all the charges are dissipated from both plates (see Figure 2-5). This time span is known as the discharging time of the capacitor.

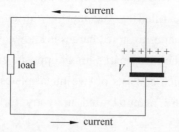

Figure 2-5　Capacitor acting as a source of electrical energy

3. How Do You Determine the Value of Capacitance?

The conducting plates have some charges q_1 and q_2 (usually, if one plate has $+q$, the other has $-q$ charge) (see Figure 2-6). The electric field in the region between the plates depends on the charge given to the conducting plates.

We also know that potential difference (V) is directly proportional to the electric field hence we can say,

$$Q \propto V$$
$$Q = CV$$

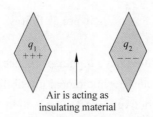

Figure 2-6　Parallel plate capacitor filled with air

$$C=Q/V$$

This constant of proportionality is known as the capacitance of the capacitor.

The capacitance of any capacitor can be either fixed or variable, depending on its usage. From the equation, it may seem that "C" depends on charge and voltage. Actually, it depends on the shape and size of the capacitor and also on the insulator used between the conducting plates.

3.1　Energy stored in a capacitor

Once the opposite charges have been placed on either side of a parallel-plate capacitor, the charges can be used to work by allowing them to move towards each other through a circuit. The equation gives the total energy that can be extracted from a fully charged capacitor:

$$U=CV^2/2$$

Capacitors function a lot like rechargeable batteries. The main difference between a capacitor and a battery lies in the technique they employ to store energy. Unlike batteries, the capacitor's ability to store energy doesn't come from chemical reactions but from the physical design that allows it to hold negative and positive charges apart.

3.2　Standard units of capacitance

The basic unit of capacitance is Farad. But Farad is a large unit for practical tasks. Hence, capacitance is usually measured in the sub-units of Farads, such as microfarads (μF) or picofarads (pF).

The standard units of capacitance are as follows:

(1) 1mF (millifarad) = 10^{-3} F

(2) 1μF (microfarad) = 10^{-6} F

(3) 1nF (nanofarad) = 10^{-9} F

(4) 1pF (picofarad) = 10^{-12} F

4. What Are the Applications of Capacitors?

4.1　Capacitors for energy storage

Since the late 18th century, capacitors have been used to store electrical energy. Individual capacitors do not hold much energy. They only provide enough power for electronic devices during temporary power outages or when they need additional power. Many applications use capacitors as energy sources, and a few of them are as follows: audio equipment, camera flashes, power supplies, magnetic coils.

Supercapacitors are capacitors that have high capacitance up to 2 kF. These capacitors store large amounts of energy and offer new technological possibilities in areas such as electric cars, regenerative braking in the automotive industry and industrial electrical motors, computer memory backup during power loss, and many others.

4.2　Capacitors for power conditioning

One of the important applications of capacitors is the conditioning of power supplies. Capacitors allow only AC signals to pass when they are charged. This capacitor effect is used in separating or decoupling different parts of electrical circuits to reduce noise as a result of improving efficiency. Capacitors are also used in utility substations to counteract inductive loading introduced by transmission lines.

4.3　Capacitors as sensors

Capacitors are used as sensors to measure a variety of things including humidity, mechanical strain, and fuel levels. Two aspects of capacitor construction are used in the sensing application — the distance between the parallel plates and the material between them. The former detects mechanical changes such as acceleration and pressure, and the latter is used in sensing air humidity.

4.4　Capacitors for signal processing

There are advanced applications of capacitors in information technology. Capacitors are used by dynamic random access memory (DRAM) devices to represent binary information as bits. Capacitors are also used in conjunction with inductors to tune circuits to particular frequencies, an effect exploited by radio receivers, speakers, and analog equalizers.

5. Frequently Asked Questions on Capacitors and Capacitance

Q1: What is a variable capacitor?

A variable capacitor is a capacitor whose capacitance can be varied to a certain range of values based on necessity. The two plates of the variable capacitor are made of metals where one of the plates is fixed, and the other is movable. Their main function is to fix the resonant frequency in the LC circuit. There are two main types of variable capacitors and they are tuning capacitors and trimming capacitors.

Q2: How does the shape of the capacitor affect its capacitance?

The distance between the plates: the more distant the plates are, the less the free electrons on the far plate feel the push of the electrons that are being added to the negative plate. This makes it harder to add more negative charges to the negative plate. The current will flow through a short circuit if the plates are closer. This implies that the capacitance of a parallel plate is inversely related to the plate separation.

Area of the plates: it's easier to add charges to a capacitor if the parallel plates have a huge area. Two wide metal plates would give two repelling-like charges a greater range to spread out across the plate, making it easier to add a lot more negative charges to one plate. Likewise, a very small plate area would cause the electrons to get cramped together earlier, making it harder to get a large difference in charge for a given voltage.

Q3: What are ultracapacitors?

An ultracapacitor, also known as the supercapacitor, is a high-capacity capacitor with a capacitance value much higher than other capacitors but with lower voltage limits.

Q4: How long does a capacitor last?

Capacitors have a limited life span. Most capacitors are designed to last approximately 20 years.

Q5: What kind of energy is stored in a capacitor?

Energy stored in a capacitor is electrical potential energy, thus related to the charge Q and voltage V on the capacitor.

Q6: Why isn't water used as a dielectric in a capacitor?

Water has a high dielectric constant but a very low dielectric strength, hence it would act as a conductor and leak charges through it.

New Words

dielectric	[ˌdaɪɪˈlektrɪk]	n.	电介质，绝缘体
gap	[gæp]	n.	间隙，间隔
girth	[gɜːθ]	n.	周长
demonstration	[ˌdemənˈstreɪʃn]	n.	演示，示范；证明
steady-state	[ˈstedɪ steɪt]	adj.	稳态的
separation	[ˌsepəˈreɪʃn]	n.	分离，分开；间隔
accumulate	[əˈkjuːmjəleɪt]	v.	堆积，积累
capacitance	[kəˈpæsɪtəns]	n.	电容
rechargeable	[rɪˈtʃɑrdʒəbl]	adj.	可再充电的
Farad	[ˈfæræd]	n.	法拉（电容单位）
micro-farad	[ˈmaɪkrəʊ ˈfærəd]	n.	微法
picofarad	[pɪkəˈfærəd]	n.	皮（可）法拉
millifarad	[ˈmɪlɪˌfærəd]	n.	毫法
nanofarad	[ˈneɪnəˈfærəd]	n.	毫微法，纳法
supercapacitor	[suːpəkəˈpæsɪtə]	n.	超级电容器
regenerative	[rɪˈdʒenərətɪv]	adj.	恢复的；再生的；新生的
separate	[ˈsepəreɪt]	v.	（使）分开，分离
decouple	[diːˈkʌpl]	vt.	去耦，解耦
counteract	[ˌkaʊntərˈækt]	vt.	抵消；阻碍；中和

fuel	[ˈfjuːəl]	n. 燃料
sense	[sens]	vt. 感知，检测到
		n. 感觉；识别力
equalizer	[ˈiːkwəlaɪzə]	n. 均衡器
fix	[fɪks]	vt. 固定
movable	[ˈmuːvəbl]	adj. 可移动的，活动的
resonant	[ˈrezənənt]	adj. 谐振的，共振的
ultracapacitor	[ˈʌltrəkəˈpæsɪtə]	n. 超电容，超级电容
approximately	[əˈprɒksɪmətlɪ]	adv. 近似地，大约
leak	[liːk]	vi. 泄密；漏电

Phrases

electronic device	电子器件，电子设备
convert…into	把……转换为
chemical energy	化学能
insulating material	绝缘材料
metal plate	金属板
parallel plate capacitor	平板电容器
positive end	正端
negative end	负端
charging time	充电时间
discharging time	放电时间
conducting plate	导电板
potential difference	电位差，电势差
be extracted from	从……提取
chemical reaction	化学反应
temporary power outage	临时停电
magnetic coil	电磁线圈
mechanical strain	机械应变
air humidity	空气湿度
binary information	二进制信息
tune circuit	调谐电路
frequently asked question	常见问题
variable capacitor	可变电容器
be varied to	变为
resonant frequency	谐振频率，共振频率
LC circuit	振荡电路
tuning capacitor	调谐电容器
trimming capacitor	微调电容器；补偿电容器

voltage limit 低电压限制
dielectric constant 介电常数，介质常数

Abbreviations

DRAM (Dynamic Random Access Memory) 动态随机存取存储器

Exercises

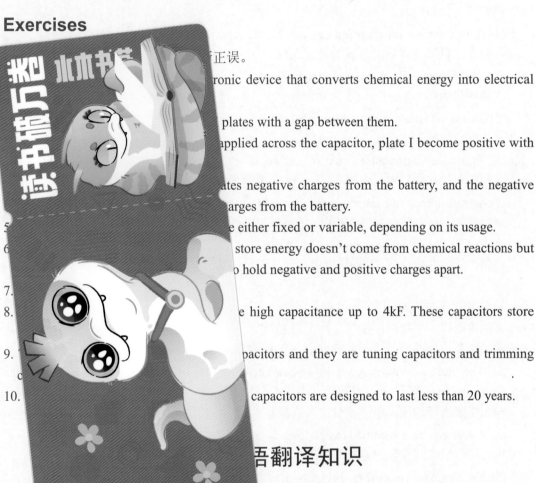

正误。

ronic device that converts chemical energy into electrical

plates with a gap between them.

applied across the capacitor, plate I become positive with

ates negative charges from the battery, and the negative
arges from the battery.

5. ... either fixed or variable, depending on its usage.
6. ... store energy doesn't come from chemical reactions but
   ... to hold negative and positive charges apart.
7.
8. ... high capacitance up to 4kF. These capacitors store
9. ... pacitors and they are tuning capacitors and trimming

10. ... capacitors are designed to last less than 20 years.

吾翻译知识

词义

在 ... 词义的问题，也就是词义选择的问题。英汉两
种语言 ... 多义的现象。一词多类指一个词往往属于几个词类，具有几个
不同的意思。一词多义是一个词在同一个词类中，又往往有几个不同的词义。词义的选择
可以从以下两方面着手。

1. 根据词类选择词义

词性不同，所代表的词义往往也不同。当译者遇到某个词，首先要弄清这个词在句子

中的词性，最后才能确定它的词义。例如：

(1) An electron is an extremely small corpuscle with negative charge which **rounds** about the nucleus of an atom.

电子是绕着原子转动带有负电荷的极其微小的粒子。（round 为动词）

(2) The earth goes **round** the sun.

地球环绕太阳运动。（round 为介词）

(3) **Round** surface reflector is a key unit for the solar energy device.

弯曲面反射器是太阳能装置的关键元件。（round 为形容词）

(4) The tree measures about one meter **round**.

这棵树树围约一米。（round 为副词）

(5) This is the whole **round** of knowledge.

这就是全部的知识范围。（round 为名词）

(6) Plastic was at first **based** on coal and wool.

最初塑料是从煤和木材中提取的。（base 为动词）

(7) As we all know, a **base** reacts with an acid to form a salt.

众所周知，碱与酸起反应变成盐。（base 为名词）

(8) Iron and brass are **base** metals.

铁和黄铜为非贵金属。（base 为形容词）

2. 根据上下文选择词义

英语中的同一个单词往往包含几个意思，在不同主题的文章中，表达的意思可能不同。因此在确定某个词的词性后，要根据上下文并结合专业知识来确定它的词义。例如：

(1) The electronic microscope possesses very high resolving **power** compared with the optical microscope.

与光学显微镜相比，电子显微镜具有极高的分辨率。

(2) **Power** can be transmitted over a long distance.

电力可以输送到很远的地方。

(3) The fourth **power** of three is eighty-one.

3 的 4 次方是 81。

(4) The combining **power** of one element in the compound must equal the combining power of the other element.

化合物中一种元素的化合价必须等于另一个元素的化合价。

(5) The medical profession enormous **power** to fight disease and sickness has been given by the explosive technological development since 1940.

1940 年以来，随着技术的迅速发展，医学界大大提高了战胜疾病的能力。

(6) Semiconductor devices have no filament or heaters and therefore require no heating **power** or warmed up time.

半导体器件没有灯丝和加热器，因此不需要加热功率或时间。

(7) A car needs a lot of **power** to go fast.

汽车高速行驶需要很大的动力。

(8) Stream and waterfall are suitable for the development of hydroelectric **power**.
溪流和瀑布适合用于开发水电能源。

Reading Material

阅读下列文章。

Text	Note
Inductor	
An inductor[1], also called a coil[2] or reactor[3], is a passive twoterminal electrical component that stores electrical energy in a magnetic field when electric current is flowing through it.	[1] *n.* 电感应器 [2] *n.* 线圈 [3] *n.* 反应器，电抗器
When the current flowing through an inductor changes, the time-varying magnetic field[4] induces a voltage in the conductor, described by Faraday's law[5] of induction. According to Lenz's law[6], the direction of induced electromotive force (e.m.f.)[7] opposes the change in current that creates it. As a result, inductors oppose any changes in current.	[4] magnetic field: 磁场 [5] Faraday's law: 法拉第定律 [6] Lenz's law: 楞次定律 [7] induced electromotive force: 感应电动势
An inductor typically consists of an electric conductor, such as a wire, that is wound into a coil.	
An inductor is characterized by its inductance, which is the ratio of the voltage to the rate of change of current. In the International System of Units (SI), the unit of inductance is the henry (H). Inductors have values that typically range from 1μH (10^{-6}H) to 1H. Many inductors have a magnetic core[8] made of iron or ferrite inside the coil, which serves to increase the magnetic field and thus the inductance[9]. Along with capacitors and resistors, inductors are one of the three passive linear[10] circuit elements that make up electric circuits. Inductors are widely used in alternating current (AC) electronic equipment, particularly in radio equipment. They are used to block AC while allowing DC to pass; inductors designed for this purpose are called chokes[11]. They are also used in electronic filters[12] to separate signals of different frequencies, and in combination with capacitors to make tuned circuits, used to tune radio and TV receivers.	[8] magnetic core: 磁芯 [9] *n.* 感应系数，自感应 [10] *adj.* 线性的 [11] *n.* 电感扼流圈 [12] *n.* 滤波器，过滤器
1. Overview	
Inductance (*L*) results from the magnetic field around a	

current-carrying conductor; the electric current through the conductor creates a magnetic flux[13]. Mathematically speaking, inductance is determined by how much magnetic flux φ through the circuit is created by a given current i

$$L=\varphi/i \qquad (1)$$

Inductors that have ferromagnetic cores are nonlinear[14]; the inductance changes with the current, in this more general case inductance is defined as

$$L=\mathrm{d}\varphi/\mathrm{d}i$$

Any wire or other conductor will generate a magnetic field when current flows through it, so every conductor has some inductance. The inductance of a circuit depends on the geometry[15] of the current path as well as the magnetic permeability[16] of nearby materials. An inductor is a component consisting of a wire or other conductor shaped to increase the magnetic flux through the circuit, usually in the shape of[17] a coil or helix. Winding the wire into a coil increases the number of times the magnetic flux lines link the circuit, increasing the field and thus the inductance. The more turns, the higher the inductance. The inductance also depends on the shape of the coil, separation of the turns, and many other factors[18]. By adding a "magnetic core" made of a ferromagnetic material like iron inside the coil, the magnetizing field from the coil will induce magnetization in the material, increasing the magnetic flux. The high permeability[19] of a ferromagnetic core can increase the inductance of a coil by a factor of several thousand over what it would be without it.

1.1 Constitutive equation

Any change in the current through an inductor creates a changing flux, inducing a voltage across the inductor. By Faraday's law of induction, the voltage induced by any change in magnetic flux through the circuit is

$$v=\mathrm{d}\varphi/\mathrm{d}t$$

From (1) above

$$v=L(\mathrm{d}i/\mathrm{d}t) \qquad (2)$$

So inductance is also a measure of the amount of electromotive force[20] (voltage) generated for a given rate of change of current. For example, an inductor with an inductance of 1 henry produces an EMF of 1 volt when the current through the inductor changes at the rate of 1 ampere[21] per second. This is usually taken to be the

[13] magnetic flux: 磁通量

[14] *adj.* 非线性的

[15] *n.* 几何形状
[16] magnetic permeability: 磁导率

[17] in the shape of: 以……的形式

[18] *n.* 因素，要素，因数

[19] *n.* 渗透性

[20] electromotive force: 电动势

[21] *n.* 安培

constitutive[22] relation (defining equation[23]) of the inductor.

The dual of the inductor is the capacitor, which stores energy in an electric field rather than a magnetic field. Its current-voltage relation is obtained by exchanging current and voltage in the inductor equations and replacing L with the capacitance[24] C.

1.2 Lenz's law

The polarity (direction) of the induced voltage is given by Lenz's law, which states that it will be such as to oppose the change in current. For example, if the current through an inductor is increasing, the induced voltage[25] will be positive at the terminal through which the current enters and negative at the terminal through which it leaves, tending to oppose the additional current. The energy from the external circuit necessary to overcome this potential "hill" is being stored in the magnetic field of the inductor; the inductor is said to be "charging" or "energizing". If the current is decreasing, the induced voltage will be negative at the terminal through which the current enters and positive at the terminal through which it leaves, tending to maintain the current. Energy from the magnetic field is being returned to the circuit; the inductor is said to be "discharging[26]".

1.3 Ideal and real inductors

In circuit theory, inductors are idealized as obeying the mathematical relation (2) above precisely. An "ideal inductor" has inductance, but no resistance or capacitance, and does not dissipate or radiate energy. However, real inductors have side effects which cause their behavior to depart from[27] this simple model. They have resistance (due to the resistance of the wire and energy losses in core material), and parasitic capacitance[28] (due to the electric field between the turns of wire which are at slightly different potentials). At high frequencies the capacitance begins to affect the inductor's behavior; at some frequency, real inductors behave as resonant circuits[29], becoming self-resonant[30]. Above the resonant frequency the capacitive reactance becomes the dominant part of the impedance. At higher frequencies, resistive losses in the windings increase due to skin effect[31] and proximity effect[32].

Inductors with ferromagnetic cores have additional energy losses due to hysteresis[33] and eddy currents[34] in the core, which increase with frequency. At high currents, iron core inductors also

[22] *adj.* 构成的，制定的

[23] *n.* 方程式，等式

[24] *n.* 电容

[25] *vt.* 感应电压

[26] *adj.* 放电的

[27] depart from: 离开

[28] parasitic capacitance: 寄生电容

[29] resonant circuits: 谐振电路

[30] self-resonant: 自谐振

[31] skin effect: 趋肤效应

[32] proximity effect: 邻近效应

[33] *n.* 磁滞现象

[34] eddy currents: 涡电流

show gradual departure from ideal behavior due to nonlinearity caused by magnetic saturation[35] of the core. An inductor may radiate electromagnetic energy into surrounding space and circuits, and may absorb electromagnetic emissions from other circuits, causing electromagnetic interference (EMI[36]). Real-world inductor applications may consider these parasitic parameters[37] as important as the inductance.

2. Applications

Inductors are used extensively in analog circuits and signal processing[38]. Applications range from the use of large inductors in power supplies to the small inductance of the ferrite bead or torus installed around a cable to prevent radio frequency interference[39] from being transmitted down the wire. Inductors are used as the energy storage device in many switched-mode power supplies to produce DC current. The inductor supplies energy to the circuit to keep current flowing during the "off" switching periods.

An inductor connected to a capacitor forms a tuned circuit, which acts as a resonator for oscillating current[40]. Tuned circuits are widely used in radio frequency equipment such as radio transmitters and receivers, as narrow bandpass filters[41] to select a single frequency from a composite signal, and in electronic oscillators[42] to generate sinusoidal[43] signals.

Two (or more) inductors in proximity that have coupled magnetic flux (mutual inductance[44] form a transformer[45], which is a fundamental component of every electric utility power grid. The efficiency of a transformer may decrease as the frequency increases due to eddy currents in the core material and skin effect on the windings. The size of the core can be decreased at higher frequencies. For this reason, aircraft use 400 hertz alternating current rather than the usual 50 or 60 hertz, allowing a great saving in weight from the use of smaller transformers.

Inductors are also employed in electrical transmission systems, where they are used to limit switching currents and fault currents. In this field, they are more commonly referred to as reactors[46].

Because inductors have complicated side effects which cause them to depart from ideal behavior, because they can radiate electromagnetic interference (EMI), and most of all, because of their

[35] magnetic saturation: 磁饱和

[36] EMI: 电磁干扰
[37] parasitic parameters: 寄生参数

[38] signal processing: 信号处理
[39] n. 干扰；冲突；干涉

[40] oscillating current: 振荡电流
[41] narrow bandpass filters: 窄带滤波器
[42] electronic oscillators: 电子振荡器
[43] n. 正弦曲线
[44] mutual inductance: 互感系数
[45] n. 变压器

[46] n. 电抗器

bulk which prevents them from being integrated on semiconductor chips[47], the use of inductors is declining in modern electronic devices, particularly compact portable devices. Real inductors are increasingly being replaced by active circuits such as the gyrator[48] which can synthesize inductance using capacitors.
 	
3. Inductor Construction
    An inductor usually consists of a coil of conducting material, typically insulated[49] copper wire, wrapped around a core either of plastic[50] or of a ferromagnetic[51] (or ferrimagnetic[52]) material; the latter is called an "iron core" inductor. The high permeability of the ferromagnetic core increases the magnetic field and confines it closely to the inductor, thereby increasing the inductance. Low frequency inductors are constructed like transformers, with cores of electrical steel laminated to prevent eddy currents. "Soft" ferrites are widely used for cores above audio frequencies, since they do not cause the large energy losses at high frequencies that ordinary iron alloys do. Inductors come in many shapes. Most are constructed as enamel coated wire (magnet[53] wire) wrapped around a ferrite bobbin[54] with wire exposed on the outside, while some enclose[55] the wire completely in ferrite and are referred to as "shielded[56]". Some inductors have an adjustable core, which enables changing of the inductance. Inductors used to block very high frequencies are sometimes made by stringing a ferrite bead on a wire.
    Small inductors can be etched[57] directly onto a printed circuit board[58] by laying out the trace in a spiral pattern. Some such planar inductors use a planar core.
    Small value inductors can also be built on integrated circuits using the same processes that are used to make transistors[59]. Aluminium[60] interconnect is typically used, laid out in a spiral coil pattern. However, the small dimensions limit the inductance, and it is far more common to use a circuit called a "gyrator" that uses a capacitor and active components to behave similarly to an inductor.
	[47] semiconductor chip: 半导体芯片
	[48] *n.* 回转器
	
	[49] *n.* 绝缘的
	[50] *n.* 塑胶，塑料
	[51] *adj.* 铁磁的
	[52] *adj.* 亚铁磁的
	
	[53] *n.* 磁体，磁铁
	[54] *n.* 线轴，绕线筒
	[55] *vt.* 封装
	[56] *adj.* 防护的，铠装的，屏蔽了的，隔离的
	
	[57] *v.* 蚀刻
	[58] printed circuit board: 印制电路板
	
	[59] *n.* 晶体管
	[60] *n.* 铝
	    adj. 铝的

参 考 译 文

电阻和电导

电导体的电阻是测量电流通过该导体的难度。反向量是电导,并且是电流通过的容易性。在概念上,电阻与机械摩擦有些相似。电阻的 SI 单位是欧姆(Ω),而电导以西门子(S)来度量。

具有均匀横截面的物体的电阻与其电阻率和长度成正比,且与其横截面积成反比。除具有零电阻的超导体之外,所有材料都有一定量的电阻。

物体的电阻(R)被定义为其上的电压(V)与通过其的电流(I)的比率,而电导(G)是相反的:

$$R = V/I \qquad G = I/V = 1/R$$

对于多种多样的材料和条件,V 和 I 彼此成正比,因此 R 和 G 是恒定的(尽管它们可以随如温度这样的其他因素而变化)。该比例被称为欧姆定律,并且满足该定律的材料被称为欧姆材料。

在其他情况下,如二极管或电池,V 和 I 不成正比。比率 V/I 有时仍然有用,并且被称为"静态电阻",因为它对应于原点和 I-V 曲线之间的弦的反斜率。在其他情况下,导数 dV/dI 可能非常有用;这被称为"差分电阻"。

1. 说明

在液压模拟中,流过导线(或电阻器)的电流类似于流经管道的水,穿过导线的电压降类似于推动水通过管道的压力降(见图 2-1)。电导与给定压力下的流量成正比,电阻与实现给定流量所需的压力成正比。电导和电阻互为倒数。

(图略)

电压降(即电阻器两侧电压之间的差)——而不是电压本身——提供推动电流通过电阻器的驱动力。这与液压类似:管道两侧之间的压力差——而不是压力本身——决定了通过它的流量。例如,在管道上方可能存在大的水压,试图将水向下推过管道。但是在管道下方可能存在相同大的水压,试图将水通过管道推回。如果这两个压力相等,水就不会流动。

电线、电阻器或其他元件的电阻和电导主要由两个属性决定:
(1)几何(形状)。
(2)材料。

几何形状很重要,因为要让水通过长而窄的管道比通过宽而短的管道更困难。同样,长而细的铜线比短而粗的铜线有更高的电阻(更低的电导)。

材料也很重要。与相同形状和大小的干净管道相比,充满头发的管道通过的水流更少。

类似地，电子可以自由且容易地流过铜线，但是不易流过具有相同形状和尺寸的钢丝，并且它们基本上不能流过绝缘体（如橡胶），无论这些绝缘体形状如何。铜、钢和橡胶之间的差异与它们的微观结构和电子构型有关，并且通过被称为电阻率的特性来量化。

2. 导体和电阻器

电可以流过的物体被称为导体。电路中使用的有特定阻力的导电材料称为电阻器（见图 2-2）。导体由高导电性材料（例如金属，特别是铜和铝）制成。另外，电阻器由多种材料制成，这取决于期望的电阻、需要消耗的能量、精度和成本等因素。

（图略）

3. 电阻率和电导率的关系

给定物体的阻力主要取决于两个因素：其材料及形状。对于给定的材料，电阻与横截面积成反比。例如，在其他方面相同的条件下，粗铜线的电阻比细铜线的电阻低。此外，对于给定材料，电阻与长度成正比。例如，在其他方面相同的条件下，长铜线比短铜线的电阻高。因此，均匀横截面导体的电阻 R 和电导 G 可以计算为

$$R=\rho(L/A)$$
$$G=\sigma(A/L)$$

其中，L 是导体长度，以米（m）度量；A 是导体的横截面积，以平方米（m^2）度量；σ（sigma）是电导率，以西门子/米（$S \cdot m^{-1}$）度量；ρ 是材料的电阻率（也称为比电阻），以欧姆·米（$\Omega \cdot m$）度量。电阻率和电导率是比例常数，因此仅取决于线材的材料，而不是线材的几何形状。电阻率和电导率互为倒数：$\rho=1/\sigma$。电阻率度量材料抵抗电流的能力。

该公式不精确，因为它假定导体中的电流密度完全均匀，在实际情况中这并不总是真实的。然而，对于像导线这样的长细导体，该公式仍然最符合实际。

该公式不精确的另一种情况是使用于交流电（AC）时，因为趋肤效应抑制了导体中心附近的电流流动，因此，几何横截面不同于电流实际流过的有效横截面，所以电阻比预期的要高。类似地，如果承载 AC 电流的两个导体彼此相邻，则由于邻近效应，它们的电阻会增加。在商业电源频率下，这些效应对于承载大电流的大导体（如变电站中的母线或承载数百安培以上的大功率电缆）是显著的。

4. 测量电阻

用于测量电阻的仪器称为欧姆表。简单的欧姆表不能准确地测量低电阻，因为其测量引线的电阻会产生电压降，从而干扰测量，因此更准确的设备使用四端子检测。

Unit 3

Text A

Simple Electric Circuit

1. An Electric Circuit

A fundamental relationship exists between current, voltage, and resistance. A simple electric circuit consists of a voltage source, some type of load, and a conductor to allow electrons to flow between the voltage source and the load.[1] In the following circuit a battery provides the voltage source, electrical wire is used for the conductor, and a light provides the resistance (see Figure 3-1). An additional component has been added to this circuit, a switch. There must be a complete path for current to flow. If the switch is open, the path is incomplete and the light will not illuminate. Closing the switch completes the path, allowing electrons to leave the negative terminal and flow through the light to the positive terminal.

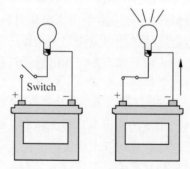

Figure 3-1   A simple electric circuit

2. An Electrical Circuit Schematic

The following schematic is a representation of an electrical circuit, consisting of a battery, a resistor, a voltmeter and an ammeter (see Figure 3-2). The ammeter, connected in series with the circuit, will show how much current flows in the circuit. The voltmeter, connected across the voltage source, will show the value of voltage supplied from the battery. Before an analysis can be made of a circuit, we need to understand Ohm's Law.

3. Ohm's Law

The relationship between current, voltage and resistance was studied by the 19th century

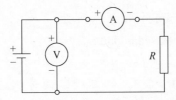

Figure 3-2 A representation of an electrical circuit

German mathematician, Georg Simon Ohm. Ohm formulated a law which states that current varies directly with voltage and inversely with resistance. From this law the following formula is derived:

$$I = \frac{U}{R} \quad \text{or} \quad \text{current} = \frac{\text{Voltage}}{\text{Resistance}}$$

Ohm's law is the basic formula used in all electrical circuits. Electrical designers must decide how much voltage is needed for a given load, such as computers, clocks, lamps and motors. Decisions must be made concerning the relationship of current, voltage and resistance. All electrical design and analysis begins with Ohm's law. There are three mathematical ways to express Ohm's law. Which of the formulas is used depends on what facts are known before starting and what facts need to be known.

$$I = \frac{U}{R} \quad U = I \times R \quad R = \frac{U}{I}$$

4. Ohm's Law Triangle

There is an easy way to remember which formula to use. By arranging current, voltage and resistance in a triangle, one can quickly determine the correct formula (see Figure 3-3).

Figure 3-3 Ohm's law triangle

5. Using the Triangle

To use the triangle, cover the value you want to calculate. The remaining letters make up the formula (see Figure 3-4).[2]

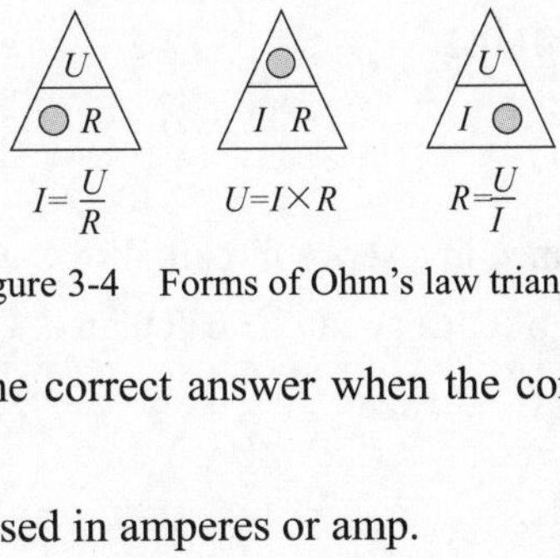

Figure 3-4 Forms of Ohm's law triangle

Ohm's law can only give the correct answer when the correct values are used. Remember the following three rules:

(1) Current is always expressed in amperes or amp.

(2) Voltage is always expressed in volt.

(3) Resistance is always expressed in ohm.

6. Resistance in a Series Circuit

A series circuit is formed when a number of resistors are connected end-to-end so that there is only one path for current to flow.[3] The resistors can be actual resistors or other devices that have resistance. The illustration shows four resistors connected end-to-end (see Figure 3-5). There is one path of electron flow from the negative terminal of the battery through R_4, R_3, R_2, R_1 returning to the positive terminal.

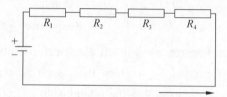

Figure 3-5　Resistance in a series circuit

7. Formula for Series Resistance

The values of resistance add in a series circuit (see Figure 3-6). If a 4Ω resistor is placed in series with a 6Ω resistor, the total value will be 10Ω. This is true when other types of resistive devices are placed in series. The mathematical formula for resistance in series is

$$R_t = R_1 + R_2 + R_3 + R_4 + R_5$$

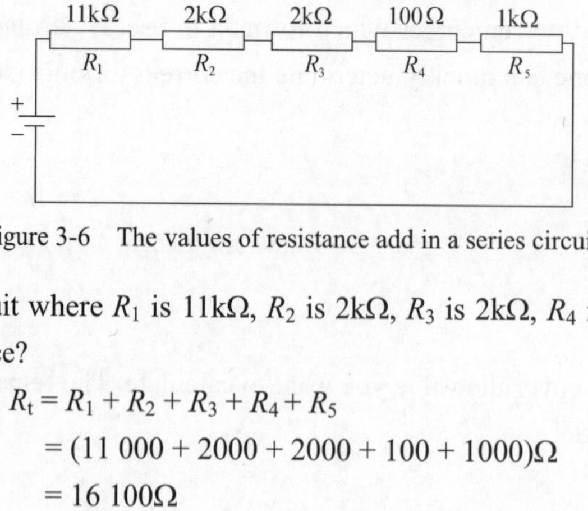

Figure 3-6　The values of resistance add in a series circuit

Given a series circuit where R_1 is 11kΩ, R_2 is 2kΩ, R_3 is 2kΩ, R_4 is 100Ω, and R_5 is 1kΩ, what is the total resistance?

$$\begin{aligned} R_t &= R_1 + R_2 + R_3 + R_4 + R_5 \\ &= (11\,000 + 2000 + 2000 + 100 + 1000)\Omega \\ &= 16\,100\Omega \end{aligned}$$

8. Current in a Series Circuit

The equation for total resistance in a series circuit allows us to simplify a circuit (see Figure 3-7). Using Ohm's law, the value of current can be calculated. Current is the same anywhere it is measured in a series circuit.

$$I = \frac{U}{R}$$

$$= \frac{12}{10} \text{A}$$
$$= 1.2 \text{A}$$

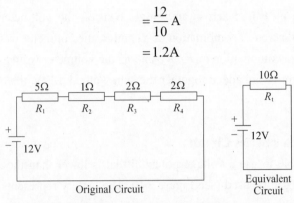

Figure 3-7　Original circuit and equivalent circuit

9. Voltage in a Series Circuit

Voltage can be measured across each of the resistors in a circuit. The voltage across a resistor is referred to as a voltage drop. A German physicist, Kirchhoff, formulated a law which states the sum of the voltage drops across the resistances of a closed circuit equals the total voltage applied to the circuit.[4] In the following illustration, four equal value resistors of 1.5Ω each have been placed in series with a 12V battery (see Figure 3-8). Ohm's law can be applied to show that each resistor will "drop" an equal amount of voltage.

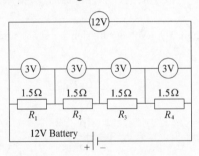

Figure 3-8　Voltage in a series circuit

First, solve for total resistance:
$$R_t = R_1 + R_2 + R_3 + R_4 = (1.5 + 1.5 + 1.5 + 1.5) \ \Omega = 6\Omega$$

Second, solve for current:
$$I = \frac{U}{R}$$
$$= \frac{12}{6} \text{A}$$
$$= 2\text{A}$$

Third, solve for voltage across any resistor:
$$U = I \times R$$
$$= (2 \times 1.5)\text{V}$$
$$= 3\text{V}$$

If voltages were measured across any single resistor, the voltmeter would read 3V.[5] If voltage were measured across a combination of R_3 and R_4 the voltmeter would read 6V. If voltage were measured across a combination of R_2, R_3, and R_4 the voltmeter would read 9V. If the voltage drops of all four resistors were added together the sum would be 12V, the original supply voltage of the battery.

10. Voltage Division in a Series Circuit

It is often desirable to use a voltage potential that is lower than the supply voltage. To do this, a voltage divider can be used (see Figure 3-9). The battery represents U_I which in this case is 50V. The desired voltage is represented by U_O which mathematically works out to be 40V. To calculate this voltage, first solve for total resistance:

$$R_t = R_1 + R_2$$
$$= (5 + 20)\Omega$$
$$= 25\Omega$$

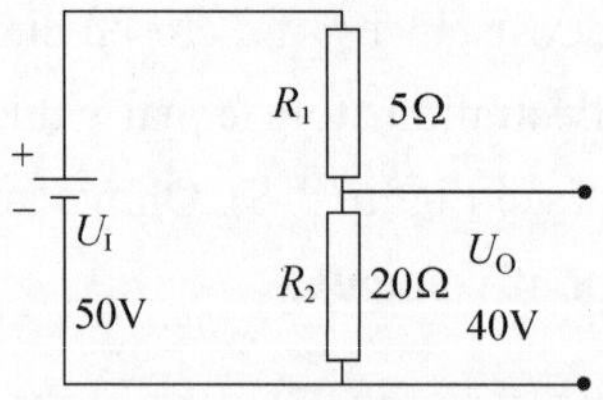

Figure 3-9 Voltage division in a series circuit

Second, solve for current:

$$I = \frac{U_I}{R_t}$$
$$= \frac{50}{25} \text{A}$$
$$= 2\text{A}$$

Finally, solve for voltage:

$$U_O = I \times R_2$$
$$= (2 \times 20)\text{V}$$
$$= 40\text{V}$$

New Words

circuit	[ˈsɜːkɪt]	n. 电路；一圈，周游，巡回
fundamental	[ˌfʌndəˈmentl]	adj. 基础的，基本的
load	[ləʊd]	n. 负荷，负载，加载
battery	[ˈbætərɪ]	n. 电池
component	[kəmˈpəʊnənt]	n. 成分
switch	[swɪtʃ]	n. 开关，电闸；转换

illuminate	[ɪˈluːmɪneɪt]	vt. 阐明，说明（问题等）；启发，启蒙
negative	[ˈnegətɪv]	adj. 阴性的，负的；否定的，消极的
		n. 否定，负数
		vt. 否定，拒绝（接受）
positive	[ˈpɒzətɪv]	adj. 阳的，带正电的；肯定的，积极的，确实的；[数]正的
schematic	[skiːˈmætɪk]	adj. 示意性的
voltmeter	[ˈvəʊltmiːtə]	n. 电压表
ammeter	[ˈæmiːtə]	n. 电流表
mathematician	[ˌmæθəməˈtɪʃn]	n. 数学家
vary	[ˈveərɪ]	vt. 改变，变更；使多样化
		vi. 变化，不同；违反
inverse	[ˌɪnˈvɜːs]	adv. 相反地，倒转地
derive	[dɪˈraɪv]	v. （使）起源于，来自；获得
designer	[dɪˈzaɪnə]	n. 设计者，设计师
lamp	[læmp]	n. 灯
relationship	[rɪˈleɪʃnʃɪp]	n. 关系，关联
analysis	[əˈnæləsɪs]	n. 分析，分解
triangle	[ˈtraɪæŋgl]	n. 三角形
calculate	[ˈkælkjuleɪt]	vt. & vi. 计算；考虑，计划，打算
		vt. & vi. （美）以为，认为
equation	[ɪˈkweɪʒn]	n. 相等，平衡；因素；方程式，等式
meter	[ˈmiːtə]	n. 仪表，计，表；米
divider	[dɪˈvaɪdə]	n. 分割者；间隔物，分配器

Phrases

electric circuit	电路
consist of	由……组成，包括，包含
be used for	用来做……，被用于
make up	组成，构成
series circuit	串联电路
series resistance	串联电阻
electron flow	电流
be placed in	被放置在
be referable to	可归因于，与……有关
voltage drop	电压降
solve for	求解
supply voltage	供电电压，电源电压
voltage potential	电压电位

Notes

[1] A simple electric circuit consists of a voltage source, some type of load, and a conductor to allow electrons to flow between the voltage source and the load.

本句中的谓语动词是 consist of，意为"由……组成，包括，包含"。to allow electrons to flow between the voltage source and the load 是一个动词不定式短语，作定语，修饰和限定 a conductor，表明是什么样的导线，而不是整个句子。

本句意为：一个简单的电路包括电源、某些类型的负载和一条让电子在电源和负载之间流动的导线。

[2] To use the triangle, cover the value you want to calculate. The remaining letters make up the formula.

这两个句子关系紧密，要联系起来理解。后一个句子表明的是使用三角形，盖住要计算的值的结果，剩下的字母组成公式。To use the triangle 是一个动词不定式短语，作目的状语，修饰 cover。you want to calculate 是一个定语从句，修饰和限定 the value。make up 的意思是"组成，构成"。

本句意为：要利用这个三角形，盖住你想要计算的值。用剩下的字符组成公式。

[3] A series circuit is formed when a number of resistors are connected end-to-end so that there is only one path for current to flow.

本句中的 end-to-end 不能凭字面理解为尾对尾，而是首尾相连的意思。so that there is only one path for current to flow 是一个结果状语从句。when a number of resistors are connected end-to-end 是一个条件状语从句。

本句意为：当多个电阻首尾相连，电流只有一条路径流动时，就形成了串联电路。

[4] A German physicist, Kirchhoff, formulated a law which states the sum of the voltage drops across the resistances of a closed circuit equals the total voltage applied to the circuit.

看懂这个句子的关键是分析它的句子结构。这是一个多层从句的句子。全句的主语是 A German physicist，谓语是 formulated，宾语是 a law，Kirchhoff 是同位语。which 引导的定语从句修饰 a law。在该定语从句中，which 作主语，states 是谓语动词，states 后又是一个宾语从句，省略了引导词 that。在这个宾语从句中，主语为 the sum of the voltage drops，谓语为 equals，宾语为 the total voltage。结构清楚后，整个句子的意思就一目了然了。

本句意为：德国物理学家基尔霍夫提出了一个定律为，整个回路中各个电阻器上的电压降的总和等于给这个回路提供的电压。

[5] If voltage were measured across any single resistor, the voltmeter would read 3V.

注意，"表的读数为……"的表达是本句中的 voltmeter would read，而不是 voltmeter would be read。read 应理解"显示，指示"。例如，The dial reads 32. 刻度显示出 32。

本句意为：如果测量任何单个电阻器上的电压，电压表的读数都会是 3V。

Exercises

【Ex.1】根据课文内容，回答以下问题。

1. What does Ohm's law state?

2. According to the passage, how to use the triangle?

3. What is a series circuit?

4. How do we measure the voltage drop of each of the resistors in a circuit?

5. If three resistors of 10Ω, 20Ω and 30Ω respectively have been placed in series with a 12V battery, what is the voltage drop of each of the resistors in a circuit?

【Ex.2】根据下面的英文解释，写出相应的英文词汇。

英 文 解 释	词 汇
a closed path followed or capable of being followed by an electric current	
a device used to break or open an electric circuit or to divert current from one conductor to another	
a position in a circuit or device at which a connection is normally established or broken	
an instrument, such as a galvanometer, for measuring potential differences in volts	
an instrument that measures electric current	
a device that generates light, heat, or therapeutic radiation	
a device that converts any form of energy into mechanical energy, especially an internal-combustion engine or an arrangement of coils and magnets that converts electric current into mechanical power	
a person who does research connected with physics or who studies physics.	
the difference in voltage between two points in an electric field or circuit	
a device that measures and records the amount of electricity, gas, water, etc. that you have used or the time and distance you have travelled, etc	

【Ex.3】把下列句子翻译成中文。
1. A power supply could be something as simple as a 9V battery or it could be as complex as a precision laboratory power supply.

2. Variable resistors have a dial or a knob that allows you to change the resistance.

3. Diodes are components that allow current to flow in only one direction.

4. LEDs use a special material which emits light when current flows through it.

5. The letter L stands for inductance. The simplest inductor consists of a piece of wire.

6. Two metallic plates separated by a non-conducting material between them make a simple capacitor.

7. The time required for a capacitor to reach its charge is proportional to the capacitance value and the resistance value.

8. When AC current flows through an inductance a back emf or voltage is generated to prevent changes in the initial current.

9. Reactance is the property of resisting or impeding the flow of AC current or AC voltage in inductors and capacitors.

10. To produce a drift of electrons, or electric current, along a wire it is necessary that there be a difference in "pressure" or potential between the two ends of the wire.

【Ex.4】把下列短文翻译成中文。

　　Switches are devices that create a short circuit or an open circuit depending on the position of the switch. For a light switch, ON means short circuit (current flows through the switch, lights light up). When the switch is OFF, that means there is an open circuit (no current flows, lights go out). When the switch is ON it looks and acts like a wire. When the switch is OFF there is no connection.

【Ex.5】通过 Internet 查找资料，借助电子词典、辅助翻译软件及 AI 工具，完成以下技术报告，并附上收集资料的网址。通过 E-mail 发送给老师，或按照教学要求在网上课堂提交。
1. 一个电路包括哪些主要元件，各种元件由哪些公司生产（附各种最新产品的图片）。
2. 叙述德国物理学家基尔霍夫的生平简历及其重大贡献。

Text B

DC Parallel Circuit

1. Resistance in a Parallel Circuit

A parallel circuit is formed when two or more resistances are placed in a circuit side-by-side so that current can flow through more than one path. The illustration shows two resistors placed side-by-side (see Figure 3-10). There are two paths of current flow. One path is from the negative terminal of the battery through R_1 returning to the positive terminal. The second path is from the negative terminal of the battery through R_2 returning to the positive terminal of the battery.

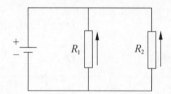

Figure 3-10　Resistance in a parallel circuit

2. Formula for Equal Value Resistors in a Parallel Circuit

To determine the total resistance when resistors are of equal value in a parallel circuit, use the following formula:

$$R_t = \frac{\text{Value of one resistor}}{\text{Number of resistors}}$$

In the following illustration there are three 15Ω resistors (see Figure 3-11). The total resistance is

$$R_t = \frac{\text{Value of one resistor}}{\text{Number of resistors}}$$
$$= \frac{15}{3} \Omega$$
$$= 5\Omega$$

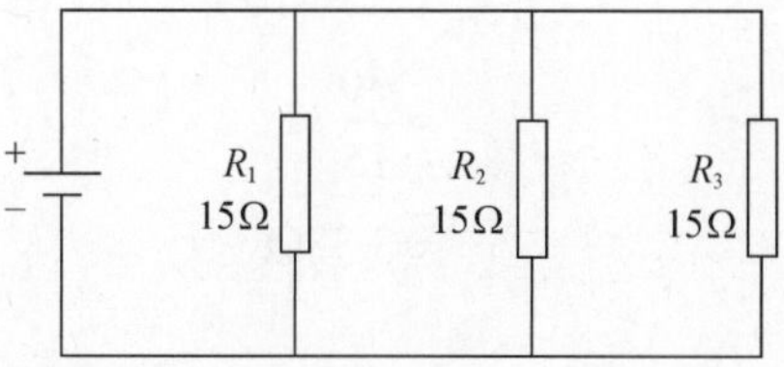

Figure 3-11　Equal value resistors in a parallel circuit

3. Formula for Unequal Resistors in a Parallel Circuit

There are two formulas to determine total resistance for unequal value resistors in a parallel circuit. The first formula is used when there are three or more resistors. The formula can be extended for any number of resistors. The following is an example of three resistors.

$$\frac{1}{R_t} = \frac{1}{R_1} + \frac{1}{R_2} + \frac{1}{R_3}$$

In the following illustration there are three resistors (see Figure 3-12). each of different value. The total resistance is

$$\frac{1}{R_t} = \frac{1}{R_1} + \frac{1}{R_2} + \frac{1}{R_3}$$

$\frac{1}{R_t} = \frac{1}{5} + \frac{1}{10} + \frac{1}{20}$ Insert value of the resistors

$= \frac{4}{20} + \frac{2}{20} + \frac{1}{20}$ Find lowest common multiple

$= \frac{7}{20}$ Add the numerators

$\frac{R_t}{1} = \frac{20}{7}$ Invert both sides of the equation

$R_t \approx 2.86\Omega$ Divide

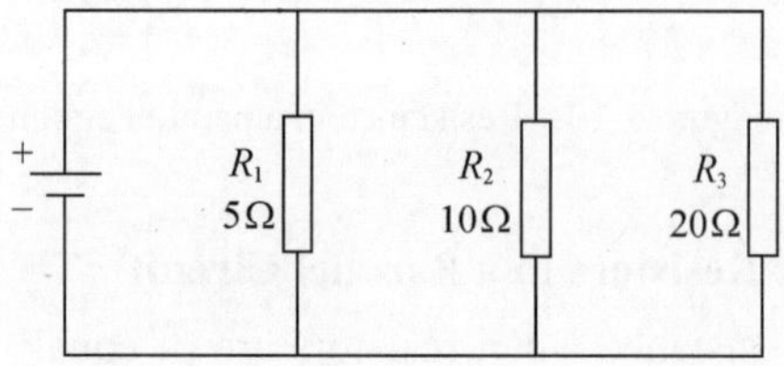

Figure 3-12 The total resistance when there are three resistors

The second formula is used when there are only two resistors.

$$R_t = \frac{R_1 \times R_2}{R_1 + R_2}$$

In the following illustration there are two resistors (see Figure 3-13), each of different value. The total resistance is

$$R_t = \frac{R_1 \times R_2}{R_1 + R_2}$$

$$= \left(\frac{5 \times 10}{5 + 10}\right) \Omega$$

$$= \frac{50}{15} \Omega$$

$$\approx 3.33\Omega$$

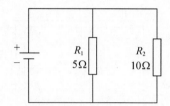

Figure 3-13 The total resistance when there are only two resistors

4. Voltage in a Parallel Circuit

When resistors are placed in parallel across a voltage source, the voltage is the same across each resistor. In the following illustration three resistors are placed in parallel across a 12V battery (see Figure 3-14). Each resistor has 12V available to it.

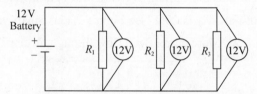

Figure 3-14 Voltage in a parallel circuit

5. Current in a Parallel Circuit

Current flowing through a parallel circuit divides and flows through each branch of the circuit (see Figure 3-15).

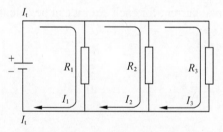

Figure 3-15 Current in a parallel circuit

Total current in a parallel circuit is equal to the sum of the current in each branch. The following formula applies to current in a parallel circuit

$$I_t = I_1 + I_2 + I_3$$

6. Current Flow with Equal Value Resistors in a Parallel Circuit

When equal resistances are placed in a parallel circuit, opposition to current flow is the same in each branch. In the following circuit R_1 and R_2 are of equal value (see Figure 3-16). If total current (I_t) is 10A, then 5A would flow through R_1 and 5A would flow through R_2.

$$\begin{aligned} I_t &= I_1 + I_2 \\ &= (5+5)\text{A} \\ &= 10\text{A} \end{aligned}$$

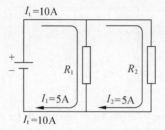

Figure 3-16 Current flow with equal value resistors in a parallel circuit

7. Current Flow with Unequal Value Resistors in a Parallel Circuit

When unequal value resistors are placed in a parallel circuit, opposition to current flow is not the same in every circuit branch. Current is greater through the path of least resistance. In the following circuit R_1 is 40Ω and R_2 is 20Ω (see Figure 3-17). Small values of resistance means less opposition to current flow. More current will flow through R_2 than R_1.

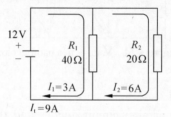

Figure 3-17 Current flow with unequal value resistors in a parallel circuit

Using Ohm's law, the total current for each circuit can be calculated.

$$I_1 = \frac{U}{R_1}$$
$$= \frac{12}{40} \text{A}$$
$$= 0.3 \text{A}$$
$$I_2 = \frac{U}{R_2}$$
$$= \frac{12}{20} \text{A}$$
$$= 0.6 \text{A}$$
$$I_t = I_1 + I_2$$
$$= (0.3 + 0.6) \text{A}$$
$$= 0.9 \text{A}$$

Total current can also be calculated by the first calculating total resistance, then applying the formula for Ohm's law.

$$R_t = \frac{R_1 \times R_2}{R_1 + R_2}$$

$$= \left(\frac{40 \times 20}{40 + 20}\right)\Omega$$

$$= \frac{800}{60}\Omega$$

$$\approx 13.333\Omega$$

$$I_t = \frac{U}{R_t}$$

$$= \frac{12}{13.333}\text{A}$$

$$\approx 0.9\text{A}$$

8. Series-parallel Circuit

Series-parallel circuit is also known as compound circuit. At least three resistors are required to form a series-parallel circuit. The following illustrations show two ways a series-parallel circuit could be found (see Figure 3-18).

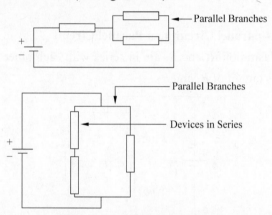

Figure 3-18　Series-parallel circuit

9. Simplifying a Series-parallel Circuit to a Series Circuit

The formulas required for solving current, voltage and resistance problems have already been defined. To solve a series-parallel circuit, reduce the compound circuits to equivalent simple circuits. In the following illustration R_1 and R_2 are parallel with each other (see Figure 3-19). R_3 is in series with the parallel circuit of R_1 and R_2.

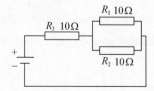

Figure 3-19　The compound circuits

First, use the formula to determine total resistance of a parallel circuit to find the total resistance of R_1 and R_2. When the resistors in a parallel circuit are equal, the following formula is used:

$$R = \frac{\text{Value of any one resistor}}{\text{Number of resistors}}$$
$$= \frac{10}{2} \Omega$$
$$= 5 \Omega$$

Second, redraw the circuit showing the equivalent values. The result is a simple series circuit which uses already learned equations and methods of problem solving (see Figure 3-20).

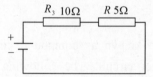

Figure 3-20　Simplifying a series-parallel circuit to a series circuit

10. Simplifying a Series-parallel Circuit to a Parallel Circuit

In the following illustration R_1 and R_2 are in series with each other (see Figure 3-21). R_3 is in parallel with the series circuit of R_1 and R_2.

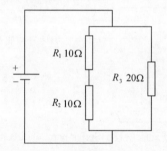

Figure 3-21　A series-parallel circuit

First, use the formula to determine total resistance of a series circuit to find the total resistance of R_1 and R_2. The following formula is used:

$$R = R_1 + R_2$$
$$= (10 + 10) \Omega$$
$$= 20 \Omega$$

Second, redraw the circuit showing the equivalent values. The result is a simple parallel circuit which uses already learned equations and methods of problem solving.

New Words

side-by-side	['saɪd baɪ saɪd]	*adj*. 并排的，并行的
multiple	['mʌltɪpl]	*n*. 倍数，若干
		adj. 多重的，多个的

numerator	['njuːməreɪtə]	n. （分数中的）分子
invert	[ɪn'vɜːt]	adj. 转化的
		vt. 使颠倒，使转化
		n. 颠倒的事物
amp	[æmp]	n. 安培
branch	[brɑːntʃ]	n. 枝，分支，支流，支脉
compound	['kɒmpaʊnd]	n. 混合物，[化]化合物
		adj. 复合的
		vt. & vi. 混合，配合
simplify	['sɪmplɪfaɪ]	vt. 单一化，简单化
reduce	[rɪ'djuːs]	vt. 减少，缩小，简化；还原
method	['meθəd]	n. 方法
redraw	[ˌriː'drɔː]	vt. 重画
		vi. 刷新（屏幕）

Phrases

parallel circuit	并联电路
negative terminal	负极端
positive terminal	正极端
lowest common multiple	最小公倍数
apply to	将……应用于
flow through	流过
be calculated by ...	用……计算
series-parallel	串-并联
compound circuits	复合电路
parallel branch	并联分支

Exercises

【Ex.6】 根据文章所提供的信息判断正误。

1. A parallel circuit is formed when two or more resistances are placed in a circuit side-by-side so that current can flow through only one path.
2. To determine the total resistance when resistors are of equal value in a parallel circuit, use the following formula

$$R_t = \frac{\text{Value of one resistor}}{\text{Number of resistors}}$$

3. In the following illustration there are three 15Ω resistors. The total resistance is 45Ω.

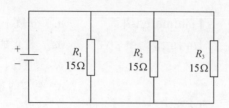

4. In the following illustration there are three resistors, each of different value. The total resistance is

$$\frac{1}{R_t} = \frac{1}{R_1} + \frac{1}{R_2} + \frac{1}{R_3}$$

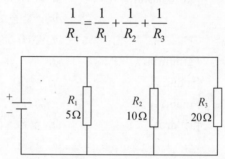

5. When resistors are placed in series across a voltage source, the voltage is the same across each resistor.
6. Current flowing through a parallel circuit divides and flows through each branch of the circuit. Total current in a parallel circuit is equal to the sum of the current in each branch.
7. When different resistances are placed in a parallel circuit, opposition to current flow is the same in each branch.
8. Series-parallel circuit is also known as compound circuits. At least more than two resistors are required to form a series-parallel circuit.
9. In the following illustration R_1 and R_2 are series with each other. R_3 is in parallel with the series circuit of R_1 and R_2.

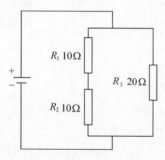

10. In the following illustration, the total resistance is 105Ω.

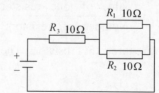

科技英语翻译知识

词义的引申

科技英语论理准确，所下的定义、定律和定理精确，所描绘的概念、叙述的生产工艺过程清楚。但是在英译汉时，经常会出现某些词在字典上找不到适当的词义的情况。如果生搬硬套，译文则生硬晦涩，不能确切表达原意，甚至有时造成误译。这时就要结合上下文，根据逻辑关系，进行词义引申，才能恰如其分地表达原意。

1. 概括化或抽象化引申

科技英语常常使用表示具体形象的词来表示抽象的意义。翻译这类词时，一般可将其词义作概括化或抽象化的引申，译文才符合汉语习惯，流畅、自然。例如：

(1) The plan for launching the man-made satellite still lies on the table.

那项发射人造卫星的计划仍被搁置，无法执行。

on the table 按字面意思译成"放在桌子上"语义不通，根据上文意思抽象引申为"无法执行"，符合原意。

(2) Military strategy may bear some similarity to the chessboard but it is dangerous to carry the analogy too far.

打仗的策略同下棋可能有某些相似之处，但是如果把这两者之间的类比搞过了头则是危险的。

chessboard 是"棋盘"。棋盘是实物，打仗的策略是思想，不好类比。因此，这里把具体的"棋盘"引申为概括性的"下棋"，就说得通了。

(3) The book is too high-powered for technician in general.

这本书对一般技术人员来说也许内容太深。

high-powered 本意为"马力大"，引申为"（艰）深"。

(4) The expense of such an instrument has discouraged its use.

这种仪器很昂贵，使其应用受到了限制。

expense 原意为花费、开支，引申为"（仪器）昂贵"。

(5) Industrialization and environmental degradation seem to go hand in hand.

工业化发展似乎伴随着环境的退化。

hand in hand 原意为"携手"，引申为"伴随"。

2. 具体化或形象化引申

科技英语中有时用代表抽象概念或属性的词来表示一种具体事物。如按字面译，则难以准确表达原文意思。这时就要根据上下文对词义加以引申，用具体或形象化的词语表达。

例如：

(1) Along the equator it reaches nearly halfway around the globe.

它沿着赤道几乎绕地球半周。

reach halfway 意为"达到一半路程"。本句讨论的是围绕地球旋转，根据这一具体语境，可以将 reach halfway 本来含义形象化地引申为"绕地球半周"。

(2) The shortest distance between raw material and a finished part is precision casting.

把原料加工成成品的最简便的方法是精密铸造。

shortest distance 原意为"最短距离"，直接按照这一字面意思，句子有失通顺。可以形象化地引申为"最简便的方法"。

(3) The foresight and coverage shown by the inventor of the process are most commendable.

这种方法的发明者所表现的远见卓识和渊博知识，给人以良好的印象。

coverage 原意为"覆盖"，引申为"渊博知识"。

(4) The purpose of a driller is to cut holes.

钻床的功能是钻孔。

purpose 原意为"目的"，引申为"功能"。

(5) There are many things that should be considered in determining cutting speed.

在测定切削速度时，应当考虑许多因素。

things 原意为"事情"，具体引申为"因素"。

Reading Material

阅读下列文章。

Text	Note

OrCAD View
 	
1. Full-Featured Schematic Editor
  OrCAD Capture, a flat[1] and hierarchical Schematic Page Editor, is based on OrCAD's legacy of fast, intuitive[2] schematic editing. Schematic Page Editor combines a standard Windows user interface with functionality and features specific to the design engineer for accomplishing design tasks and publishing design data.
  (1) Undo and redo schematic edit unlimited times.
  (2) Use Label State for "what if" scenarios[3].
  (3) Launch Property Spreadsheet Editor at design or schematic level to edit or print your design properties.
  (4) View and edit multiple schematic designs in a single session.
  (5) Reuse design data by copying and pasting within or between schematics.
	[1] *adj*. 平面的
	[2] *adj*. 直觉的
	
	[3] *n*. 情况

(6) Select parts from a comprehensive[4] set of functional part libraries.

(7) In-line editing of parts to allow pin name and number movement.

(8) File locking in case the design is being open by another user.

2. OrCAD Capture

OrCAD Capture® design entry is the most widely used schematic entry system in electronic design today for one simple reason: fast and universal design entry. Whether you're designing a new analog circuit, revising schematic diagram for an existing PCB[5], or designing a digital block diagram with an HDL module, OrCAD Capture provides simple schematic commands you need to enter, modify and verify the design for PCB.

(1) Place, move, drag, rotate, or mirror individual parts or grouped selections while preserving both visual and electrical connectivity[6].

(2) Ensure design integrity through configurable Design and Electrical Rule checkers.

(3) Create custom title blocks and drawing borders to meet the most exacting specifications.

(4) Insert drawing objects, bookmarks, logos[7] and bitmapped pictures.

(5) Choose from metric or imperial[8] unit grid spacing to meet all drawing standards.

(6) Design digital circuits with VHDL or Verilog Text Editor.

Find and select parts or nets quickly from the OrCAD Capture Project Manager and the multi-window interface makes navigation[9] across hierarchy a breeze.

3. Project Manager Coordinates Design Data

The sophisticated[10] Project Manager simplifies organizing and tracking the various types of data generated in the design process.

An expanding-tree diagram makes it easy to structure and navigate all of your design files, including those generated by PSpice® simulators, Capture CIS and other plug-ins.

(1) Project Creation Wizard guides you through all the resources available for a specific design flow.

---

[4] *adj.* 全面的，广泛的

[5] Printed Circuit Board，印制电路板

[6] *n.* 连通性

[7] *n.* 标识

[8] *adj.* 英制的（度量衡）

[9] *n.* 向导，导航

[10] *adj.* 先进的，精妙的

(2) Centralized management of all design data permits a seamless[11] interchange of schematic data for OrCAD plug-ins and downstream flow.

(3) Hierarchy browser lets you navigate the entire schematic structure and open specific elements whether it's a schematic page, a part, or net—instantly.

(4) File tab groups multi-page schematics in folders for flat designs and creates new folders automatically for added levels of hierarchical[12] designs.

(5) Archive capability ensures the portability[13] of your entire design project.

4. Hierarchical Design and Reuse

OrCAD Capture boosts[14] schematic editing efficiency by enabling you to reuse subcircuits[15] without having to make multiple copies. Instead, using hierarchical blocks, you can simply reference the same subcircuit multiple times.

(1) Enable a single instance of the circuitry[16] for you to create, duplicate and maintain.

(2) Automatic creation of hierarchical ports eliminates potential design connection errors.

(3) Update ports and pins dynamically for hierarchical blocks and underlying schematics.

(4) Reuse OrCAD Layout and Cadence® Allegro® high-speed PCB modules within or between schematics.

(5) Require just one instance of the circuitry for you to create and maintain.

(6) Unlimited referencing and reuse of circuitry throughout your entire design.

(7) Serve schematic pages from library files.

(8) Sophisticated Property Editor clearly distinguishes[17] properties in a subcircuit from those in referenced uses allowing you to view and edit from one place.

5. Libraries And Part Editor

You can access Library Editor directly from the OrCAD Capture user interface. Create and edit parts in the library or directly from the schematic page without interrupting your workflow.

---

[11] *adj.* 无缝的，无痕的

[12] *adj.* 分等级的

[13] *n.* 可携带，轻便

[14] *vt. & vi.* 推进

[15] *n.* 支电路

[16] *n.* 电路，线路

[17] *vt. & vi.* 区别，辨别

(1) Movable pin name and pin number.

(2) Intuitive graphical controls speed of schematic part creation and editing.

(3) Create new parts quickly by modifying existing ones.

(4) Spreadsheet[18] and pin array utilities make short work of creating and editing pin-intensive devices.

(5) Bused vector pins reduce clutter on schematics.

(6) Create FPGA and CPLD symbols quickly and easily with Part Generator. Compatible with ten popular places and route pin reports.

(7) Drag-and-drop parts between libraries.

(8) Speed creation and maintenance of master library sets with design cache.

(9) Revise a part in the original subcircuit only, or propagate[19] the change to all other uses of the subcircuit in the design.

(10) Capability to add or delete sections of multisection[20] homogeneous/heterogeneous parts.

(11) Control power and ground pin visibility and connectivity on a per-schematic basis.

6. Integrate Huge I/O Count FPGA And CPLD

OrCAD Capture provides a Library Part Generator to automate the integration of FPGA and PLD[21] devices into your system schematic. The Generate Part feature simplifies the creating of core FPGA library parts for devices that might have many hundreds of pins. Signal placement reports created by popular FPGA design applications like those from Altera, Actel, and Xilinx are read into Generate Part to design the core Capture symbol saving up the hours of tedious graphical entry work. OrCAD Capture supports Xilinx 4.1i/4.2iPAD file format. If, during the PCB layout phase, the PCB designer discovers a more efficient pin placement scheme for the package or additional functionality[22] is added to the FPGA or PLD, the system engineer must modify the symbol and schematics to reflect this change which is error prone and may cause designs to be out of sync. The Generate Part feature has an annotate[23] option which modifies an existing symbol with new pin assignments.

Step 1: Creating parts with potentially hundreds of pins is an error prone and painstaking task. With Generate Part you simply browse in

[18] *n.* 电子表格，电子制表软件，电子数据表

[19] *vt. & vi.* 繁殖，传播

[20] 多节，多段

[21] 可编程逻辑器件

[22] *n.* 功能性，泛函性

[23] *vt. & vi.* 注释，评注

the pin and signal report file created by your place and route software.

Step 2: Specify to create a new part or update an existing one. Packages of all kinds are supported including PGAs and BGAs.

Step 3: The new part is created fast. Pins with common names are intelligently[24] grouped and ordered.

[24] *adv.* 聪明地，智能地

7. Easy Entry Of Part, Pin, And Net Data

Access all part, net, pin, and title block properties, or any subset, and make changes quickly through the OrCAD Capture spreadsheet Property Editor.

(1) Select a circuit element, grouped area, or entire page then add/edit/delete part, net, or pin properties.

(2) Globally apply specific property names across all your designs to meet your particular netlist[25] or other output requirements. This maintains consistency, reduces manual errors, and eliminates multiple re-entry.

[25] *n.* 连线表

(3) Browse and instantly visit any part, net, hierarchical port, off page connector, bookmark, or design rule error marker from a single reference point.

8. Verify Circuits Early With Design Rule Check

The configurable Design Rule Check (DRC) feature in OrCAD Capture allows a comprehensive verification of your design before committing[26] to downstream design processes saving the time and cost of ECOs latter in the design cycle.

[26] *vt.* 把……交托给，提交

(1) Report duplicate parts.

(2) Identify invalid design packaging.

(3) Detect off-grid objects leading to unconnected signals.

(4) Configure with electrical violations to report and assign severity[27] warnings.

[27] *n.* 严格，严重

(5) Check entire design or specific modules.

9. Reports

OrCAD Capture creates basic bill of materials (BOMs) outputs extracting from the information contained in the schematic database.

(1) Extract all part properties in the schematic design and output them to a text file.

(2) Automatically package parts with reference designators prior

to report generation.

10. Part Selection

While placing a component, you can identify it visually, modify the properties as needed, then dynamically place it within a design—all in the same sequence.

(1) Zero-in quickly on the exact library part you want, using wildcard[28] searches.

(2) Pick your recent part choices from the most recently used (MRU) menu.

(3) Choose a logic gate or DeMorgan equivalent.

(4) Edit schematic parts graphically prior to placement.

(5) Add, modify, and delete part properties at any time.

(6) Place previously used parts fast by grabbing them from the project design cache.

(7) Automatically assign reference designators during or after part placement. Update all, or just unidentified[29] parts, or reset all to placeholder values.

(8) Add libraries to a project from any drive or directory without leaving the part selector.

(9) Apart filter[30] can be used to filter out the parts from existing libraries based on parameters like HDL models, Spice models, etc. associated with symbols.

11. Interface Capabilities

OrCAD Capture interfaces with other CAD applications with minimal[31] translation needs or integration problems by importing and exporting virtually every commonly used design file format.

(1) Export of DXF files to AutoCAD™.

(2) View and redline schematic with MYRIAD™.

(3) Bi-directional EDIF 200 graphic transfer and export of the EDIF 200 netlist format.

(4) Import MicroSim® schematic.

(5) Export of more than 30 netlist formats, including VHDL, Verilog®, PSpice, SPICE, and PADS 4.0.

(6) Interface with OrCAD Layout and Allegro PCB with forward and back annotation.

(7) Interface with NC VHDL Desktop and Synplicity Synplify®

[28] *n.* 通配符

[29] *adj.* 未经确认的

[30] *n.* 滤波器

[31] *adj.* 最小的，最小限度的

for FPGA design. 　　(8) Interface with NC VHDL Desktop and NC Verilog® Desktop for board level (multi-chips) digital simulation[32]. 　　(9) Creation of custom netlists using Microsoft Visual BASIC. **12. Printing and Plotting** 　　Produce professional hardcopy through any output device supported by Microsoft Windows. 　　(1) Print Area prints specific area of the design in larger scale. 　　(2) Print Preview ensures proper scale and orientation[33]. 　　(3) Export to the DXF format for CAD interchange. 　　(4) Cross-probing[34] between OrCAD Capture and Cadence Allegro PCB Layout.	[32] *n.* 仿真，模拟 [33] *n.* 方向，方位，定位 [34] *n.* 探测，探查

参 考 译 文

简 单 电 路

1. 电路

电流、电压和电阻之间存在最基本的关系。一个简单的电路包括电源、某些类型的负载和一条让电子在电源和负载之间流动的导线。在下面的电路中，电池提供电压源，电线用作导体，灯泡提供电阻，如图 3-1 所示。开关作为电路的附加元件，由此形成一个完整的电路。如果开关断开，电路不通，电灯将不亮。闭合开关，电路接通，则电子通过灯泡从负极流向正极。

（图略）

2. 电路示意图

下面是一个电路示意图，包括 1 个电池、1 个电阻、1 个电压表和 1 个电流表，如图 3-2 所示。电流表串联在电路中，用于显示有多大的电流流过电路。电压表跨接于电压源，用于显示电池提供多大的电压。在分析一个电路的组成之前，我们先了解欧姆定律。

（图略）

3. 欧姆定律

19 世纪德国数学家乔治·西蒙·欧姆研究了电流、电压和电阻之间的关系。欧姆提出了一个定律，即电流与电压成正比，与电阻成反比。从该定律可以导出下面的公式：

（公式略）

欧姆定律是适用于全部电路的基本定律。电气设计者必须决定对于给定的负载需要多大的电压，如计算机、时钟、灯和电动机。这种决定必定涉及电流、电压和电阻之间的关系。所有的电气设计和分析都是由欧姆定律开始的。有以下 3 种表达欧姆定律的数学方法。使用哪个公式取决于开始之前已知哪些事实及需要了解哪些事实。

（公式略）

4．欧姆定律三角形

有一种简单的方法来记住使用哪个公式。通过将电流、电压和电阻安排在一个三角形中，就可以迅速确定正确的公式。

（图略）

5．利用三角形

要利用这个三角形，盖住你想要计算的值。用剩下的字符组成公式。

（图略）

只有在使用正确的值时欧姆定律才能给出正确的答案。记住下面的 3 个规则：

（1）电流总是用安培表示的。

（2）电压总是用伏特表示的。

（3）电阻总是用欧姆表示的。

6．在串联电路中的电阻

当多个电阻首尾相连，电流只有一条路径流动时，就形成了串联电路。电阻可以是实际的电阻器或者是有电阻的其他设备。插图显示 4 个电阻首尾连接，如图 3-3 所示。电流有一条路径，从电池负极经 R_4、R_3、R_2、R_1 返回正极。

（图略）

7．串联电阻公式

在串联电路中阻值是相加的，如图 3-4 所示。如果 1 个 4Ω 的电阻器同 1 个 6Ω 的电阻器连在一起，总电阻值是 10Ω。其他有阻抗的设备串联也是这样的。电阻串联的数学公式是

（公式略）

给出的串联电路中，R_1 是 11kΩ，R_2 是 2kΩ，R_3 是 2kΩ，R_4 是 100Ω，R_5 是 1kΩ，电路的总电阻是多少？

（计算过程略）

8．串联电路中的电流

串联电路中总电阻的公式使我们能够简化电路，如图 3-5 所示。利用欧姆定律就可以计算出电流的值。在串联电路中的任何地方测量电流的值都是相同的。

（公式及图略）

9. 串联电路中的电压

可以通过电路中的每个电阻器来测量电压。电压经过一个电阻器就意味着一个电压降。德国物理学家基尔霍夫提出了一个定律：整个回路中各个电阻器上的电压降的总和等于给这个回路提供的电压。在下面的示意图中，4个阻值都为1.5Ω的电阻串联起来连接一个12V的电池，如图3-6所示。根据欧姆定律可以知道每个电阻器上"下降"相等的电压。

（图略）

首先，求解总电阻。

（公式及计算略）

第二，求解电流。

（公式及计算略）

第三，求解任意电阻上的电压。

（公式及计算略）

如果测量任何单个电阻器上的电压，电压表的读数都会是3V。如果测量R_3和R_4组合上的电压，电压表的读数是6V。如果测量R_2、R_3和R_4组合上的电压，电压表的读数是9V。如果将4个电阻器上的电压降加起来，其和将是给电路提供电压的电池的电压，即12V。

10. 串联电路上分压

人们往往希望使用低于供电电压的电压。为此，可以使用电压分配器，如图3-7所示。电池电压用U_I表示，本例中为50V。希望得到的电压用U_O表示。通过计算得出是40V。为了计算这个电压，首先求解总电阻。

（图、公式及计算略）

第二，求解电流。

（公式及计算略）

最后，求解电压。

（公式及计算略）

Text A

Microcontroller

A microcontroller is a compact integrated circuit designed to govern a specific operation in an embedded system. A typical microcontroller includes a processor, memory and input/output (I/O) peripherals on a single chip.

Sometimes referred to as an embedded controller or microcontroller unit (MCU), microcontrollers are found in vehicles, robots, office machines, medical devices, mobile radio transceivers, vending machines and home appliances, among other devices. They are essentially simple miniature personal computers (PCs) designed to control small features of a larger component. They do not have a complex front-end operating system (OS).

1. How Do Microcontrollers Work?

A microcontroller is embedded inside of a system to control the specific functions in a device. It does this by interpreting data it receives from its I/O peripherals using its central processor. The temporary information that the microcontroller receives is stored in its data memory, the processor accesses it and uses instructions stored in its program memory to decipher and apply the incoming data.[1] It then uses its I/O peripherals to communicate and enact the appropriate action.

Microcontrollers are used in a wide array of systems and devices. Devices often utilize multiple microcontrollers that work together within the device to handle their respective tasks.

For example, a car might have many microcontrollers that control various individual systems within, such as the anti-lock braking system, traction control, fuel injection or suspension control. All the microcontrollers communicate with each other to inform the correct actions. Some might communicate with a more complex central computer within the car, and others might only communicate with other microcontrollers. They send and receive data using their I/O peripherals and process that data to perform their designated tasks.

2. What Are the Elements of a Microcontroller?

2.1 The processor (CPU)

A processor can be thought of as the brain of the device. It processes and responds to various instructions that direct the microcontroller's function. This involves performing basic arithmetic, logic and I/O operations. It also performs data transfer operations and communicates commands to other components in the larger embedded system.

2.2 Memory

A microcontroller's memory is used to store the data that the processor receives and uses to respond to instructions that it's been programmed to carry out.[2] A microcontroller has two main memory types:

(1) Program memory, which stores long-term information about the instructions that the CPU carries out. Program memory is non-volatile memory, meaning it holds information over time without needing a power source.

(2) Data memory, which is required for temporary data storage while the instructions are being executed. Data memory is volatile, meaning the data it holds is temporary and is only maintained if the device is connected to a power source.

2.3 I/O peripherals

The input and output devices are the interface for the processor to the outside world. The input ports receive information and send it to the processor in the form of binary data. The processor receives that data and sends the necessary instructions to output devices that execute tasks external to the microcontroller.

While the processor, memory and I/O peripherals are the defining elements of the microprocessor, there are other elements that are frequently included. The term I/O peripherals itself simply refers to supporting components that interface with the memory and processor. There are many supporting components that can be classified as peripherals.

2.4 Analog to digital converter (ADC)

An ADC is a circuit that converts analog signals to digital signals. It allows the processor at the center of the microcontroller to interface with external analog devices, such as sensors.

2.5 Digital to analog converter (DAC)

A DAC performs the inverse function of an ADC and allows the processor at the center of the microcontroller to communicate its outgoing signals to external analog components.

2.6 System bus

The system bus is the connective wire that links all components of the microcontroller together.

2.7 Serial port

The serial port is one example of an I/O port that allows the microcontroller to connect to external components. It has a similar function to a USB or a parallel port but differs in the way it exchanges bits.

3. Features of Microcontrollers

A microcontroller's processor will vary by application. Options range from the simple 4-bit, 8-bit or 16-bit processors to more complex 32-bit or 64-bit processors. Microcontrollers can use volatile memory types and non-volatile memory types. Volatile memory types include random access memory (RAM), and non-volatile memory types include flash memory, erasable programmable read-only memory (EPROM) and electrically erasable programmable read-only memory (EEPROM).

Generally, microcontrollers are designed to be readily usable without additional computing components because they are designed with sufficient onboard memory as well as offering pins for general I/O operations, so they can directly interface with sensors and other components.[3]

Microcontroller architecture can be based on the Harvard architecture or von Neumann architecture, both offering different methods of exchanging data between the processor and memory. With a Harvard architecture, the data bus and instruction are separate, allowing for simultaneous transfers. With a von Neumann architecture, one bus is used for both data and instructions.

Microcontroller processors can be based on complex instruction set computing (CISC) or reduced instruction set computing (RISC). While CISC is easier to implement and more efficient in memory use, it may lead to performance degradation due to the higher number of clock cycles needed to execute instructions.[4] RISC processors simplify the instruction set, thus increasing design simplicity.

When they first became available, microcontrollers solely used the assembly language. Today, the C programming language is a popular option. Other common microprocessor languages include Python and JavaScript.

MCUs feature input and output pins to implement peripheral functions. Such functions include analog to digital converters, liquid crystal display (LCD) controllers, real-time clock (RTC), universal synchronous/asynchronous receiver transmitter (USART), timers, universal asynchronous receiver transmitter (UART) and universal serial bus (USB) connectivity.

4. Commonly Used Microcontrollers

(1) Intel MCS-51, often referred to as an 8051 microcontroller, which was first developed in 1985;

(2) AVR microcontroller, developed by Atmel in 1996;

(3) Programmable interface controller (PIC) from Microchip Technology; and

(4) Various licensed Advanced RISC Machines (ARM) microcontrollers.

5. Applications of Microcontrollers

Microcontrollers are used in multiple industries and applications, including in the home and enterprise, building automation, manufacturing, robotics, automotive, lighting, smart energy, industrial automation, communications and internet of things (IoT) deployments.

One very specific application of a microcontroller is its use as a digital signal processor. Frequently, incoming analog signals come with a certain level of noise. Noise in this context means ambiguous values that cannot be readily translated into standard digital values.[5] A microcontroller can use its ADC and DAC to convert the incoming noisy analog signal into a clean outgoing digital signal.

The simplest microcontrollers facilitate the operation of electromechanical systems found in everyday convenience items, such as ovens, refrigerators, toasters, mobile devices, key fobs, video game systems, televisions and lawn-watering systems. They are also common in office machines such as photocopiers, scanners, fax machines and printers, as well as smart meters, ATMs and security systems.

More sophisticated microcontrollers perform critical functions in aircraft, spacecraft, ocean-going vessels, vehicles, medical and life-support systems as well as in robots.

6. Factors Influencing Microcontroller Selection

When choosing a microcontroller for a particular application, several factors come into play:

(1) Processing power: the complexity of the task dictates the required processing power, which influences the choice between 8-bit, 16-bit, or 32-bit microcontrollers.

(2) Memory capacity: sufficient memory is essential for storing program code, data, and variables. More memory is needed for applications with larger codebases or data storage requirements.

(3) Peripheral support: depending on the application, required peripherals such as ADCs, DACs, timers, communication interfaces must be considered.

(4) Power efficiency: battery-powered devices require microcontrollers optimized for low power consumption to prolong battery life.

(5) Cost: project budget constraints may influence the choice of microcontrollers, as higher-performance microcontrollers tend to be more expensive.

New Words

microcontroller	[maɪkrəʊkɒn'trəʊlər]	n. 微控制器
compact	[kəm'pækt]	adj. 紧凑的；简洁的
peripheral	[pə'rɪfərəl]	n. 外部设备
transceiver	[træn'siːvə]	n. 收发机，收发器
front-end	[frʌnt end]	adj. 前端的
embed	[ɪm'bed]	v. 把……嵌入
temporary	['temprərɪ]	adj. 临时的，暂时的

instruction	[ɪn'strʌkʃn]	n.	指令
decipher	[dɪ'saɪfə]	vt.	破译（密码）；解读
		n.	密电译文
respective	[rɪ'spektɪv]	adj.	各自的，分别的
respond	[rɪ'spɒnd]	v.	响应；回答
arithmetic	[ə'rɪθmətɪk]	n.	算术，计算
non-volatile	['nʌn,vɒlətaɪl]	adj.	非易失性的
hold	[həʊld]	vt.	保留，保存
execute	['eksɪkjuːt]	vt.	执行，完成
maintain	[meɪn'teɪn]	vt.	保持；保养
bus	[bʌs]	n.	总线
link	[lɪŋk]	n. & v.	链接；关联
option	['ɒpʃn]	n.	选择，选项，选择权
separate	['sepəreɪt]	v.	分开，分离，分散
simultaneous	[,sɪml'teɪnɪəs]	adj.	同时的
transfer	[træns'fɜː]	v.	传输，转移
	['trænsfɜː]	n.	传输，转移
performance	[pə'fɔːməns]	n.	性能，表现
degradation	[,degrə'deɪʃn]	n.	恶化，下降
robotic	[rəʊ'bɒtɪk]	adj.	机器人的；自动的
deployment	[dɪ'plɔɪmənt]	n.	部署；调度
ambiguous	[æm'bɪgjuəs]	adj.	含糊的，不明确的
standard	['stændəd]	n.	标准，规格
		adj.	标准的
electromechanical	[ɪ'lektrəʊmɪ'kænɪkəl]	adj.	电动机械的，机电的
photocopier	['fəʊtəʊkɒpɪə]	n.	复印机
scanner	['skænə]	n.	扫描仪，扫描设备
printer	['prɪntə]	n.	打印机；印刷机
dictate	[dɪk'teɪt]	vt.	控制，支配
variable	['veərɪəbl]	n.	变量
codebase	[kəʊdbeɪs]	n.	代码库
prolong	[prə'lɒŋ]	vt.	延长，拉长

Phrases

single chip	单片
vending machine	自动售货机
anti-lock braking system (ABS)	防抱装置
communicate with ...	与……通信；与……联系
carry out	执行，进行；完成

long-term information	长期信息
be classified as	被归类为
analog device	模拟设备，模拟装置
system bus	系统总线
serial port	串行端口
external component	外部元件，外部部件
volatile memory	易失性存储器
non-volatile memory	非易失性存储器
flash memory	闪存
onboard memory	板载存储器
be based on	基于
von Neumann architecture	冯·诺依曼体系结构
data bus	数据总线
clock cycle	时钟（脉冲）周期
instruction set	指令集
assembly language	汇编语言
programming language	编程语言
building automation	楼宇自动化
smart energy	智能能源
be translated into	被转换为
key fob	密钥卡；遥控钥匙
video game system	视频游戏系统
fax machine	传真机
smart meter	智能电表
ocean-going vessel	远洋船舶
program code	程序代码
battery-powered device	电池供电装置
project budget	项目预算

Abbreviations

MCU (Microcontroller Unit)	微控制器单元
PC (Personal Computer)	个人计算机
OS (Operating System)	操作系统
USB (Universal Serial Bus)	通用串行总线
RAM (Random Access Memory)	随机存储器
EPROM (Erasable Programmable Read-Only Memory)	可擦可编程只读存储器
EEPROM (Electrically Erasable Programmable Read-Only Memory)	电擦除可编程只读存储器
CISC (Complex Instruction Set Computing)	复杂指令集计算技术

RISC (Reduced Instruction Set Computing)　　精简指令集计算技术
LCD (Liquid Crystal Display)　　液晶显示器
RTC (Real-Time Clock)　　实时时钟
USART (Universal Synchronous/Asynchronous Receiver Transmitter)　　通用同步/异步接收发送设备
UART (Universal Asynchronous Receiver Transmitter)　　通用异步接收发送设备
PIC (Programmable Interface Controller)　　可编程接口控制器
ARM (Advanced RISC Machine)　　高级精简指令集计算机
IoT (internet of things)　　物联网
ATM (Automated Teller Machine)　　自动柜员机，自动取款机

Notes

[1] The temporary information that the microcontroller receives is stored in its data memory, the processor accesses it and uses instructions stored in its program memory to decipher and apply the incoming data.

本句中，主语是 The temporary information，谓语是 is stored，in its data memory 是地点状语。that the microcontroller receives 是一个定语从句，修饰和限定主语。stored in its program memory 是一个过去分词短语作定语，修饰和限定 instructions。to decipher and apply the incoming data 是一个动词不定式短语，作目的状语，修饰 uses。

本句意为：微控制器接收到的临时信息存储在其数据存储器中，处理器访问该信息并使用存储在程序存储器中的指令来解读和应用输入的数据。

[2] A microcontroller's memory is used to store the data that the processor receives and uses to respond to instructions that it's been programmed to carry out.

本句中，to store the data that the processor receives and uses 是一个动词不定式短语，作目的状语，修饰主句的谓语 is used。在该短语中，that the processor receives and uses 是一个定语从句，修饰和限定 the data。to respond to instructions that it's been programmed to carry out 是一个动词不定式短语，作目的状语，修饰 to store。that it's been programmed to carry out 是一个定语从句，修饰和限定 instructions。

本句意为：微控制器的存储器用于存储处理器接收并使用的数据，来响应编程的要执行的指令。

[3] Generally, microcontrollers are designed to be readily usable without additional computing components because they are designed with sufficient onboard memory as well as offering pins for general I/O operations, so they can directly interface with sensors and other components.

本句中，because they are designed with sufficient onboard memory as well as offering pins for general I/O operations, so they can directly interface with sensors and other components 是一个原因状语从句，修饰主句的谓语。在该从句中，so they can directly interface with sensors and other components 是一个目的状语从句，修饰从句的谓语。

本句意为：一般来说，微控制器被设计为易于使用，无须额外的计算部件，因为它们

被设计有足够的板载内存,并提供用于一般 I/O 操作的引脚,所以它们可以直接与传感器和其他部件连接。

[4] While CISC is easier to implement and more efficient in memory use, it may lead to performance degradation due to the higher number of clock cycles needed to execute instructions.

本句中,While CISC is easier to implement and more efficient in memory use 是一个让步状语从句,修饰主句的谓语。due to the higher number of clock cycles needed to execute instructions 是一个原因状语,也修饰主句的谓语。needed to execute instructions 是一个过去分词短语,作定语,修饰和限定 the higher number of clock cycles。

本句意为:虽然 CISC 更容易实现并且内存使用效率更高,但由于执行指令所需的时钟周期数较多,它可能会导致性能下降。

[5] Noise in this context means ambiguous values that cannot be readily translated into standard digital values.

本句中,in this context 是一个介词短语,作定语,修饰和限定 Noise。that cannot be readily translated into standard digital values 是一个定语从句,修饰和限定 ambiguous values。

本句意为:在这种情况下,噪声意味着不明确的值,它们无法轻易转换为标准数字值。

Exercises

【Ex.1】根据课文内容,回答以下问题。

1. What is a microcontroller? What does a typical microcontroller include?

2. What can a processor be thought of? What does it do?

3. What can microcontroller architecture be based on? What do they offer?

4. Are microcontrollers used in multiple industries and applications? What do they include?

5. What are the several factors that come into play when choosing a microcontroller for a particular application?

【Ex.2】根据下面的英文解释，写出相应的英文词汇。

英 文 解 释	词　汇
a small computer on a single integrated circuit, which contains one or more CPUs (processor cores) along with memory and programmable input/output peripherals	
an electronic equipment connected by cable to the CPU of a computer	
an electronic device which is a combination of a radio transmitter and a receiver	
a piece of information that tells a computer to perform a particular operation	
relating to or consisting of calculations involving numbers	
a set of wires that carries information from one part of a computer system to another	
a link of a network is one of the connections between the nodes of the network	
the act of moving something from one place to another	
a device that captures images from photographic prints, posters, magazine pages and similar sources for computer editing and display	
a machine for printing text or pictures onto paper, especially one linked to a computer	

【Ex.3】把下列句子翻译成中文。

1. A computer system includes a computer, peripheral devices, and software.

2. To visit similar websites to this one, click on the links at the bottom of the page.

3. The printer are linked to a powerful computer.

4. This printer is compatible with most microcomputers.

5. Scanning speed is an important performance index of a scanner.

6. Have you taken all the variables into account in your calculations?

7. They set high standards of customer service.

8. Select the landscape option when printing the file.

9. The non-volatile memory has advantages such as low leakage power and high density.

10. I'm still no closer to deciphering the code.

【Ex.4】把下列短文翻译成中文。

What Is an Operating System?

An operating system is a software interface between the device hardware and the user. It enables users to interact with the device and conduct their desired functions. OS supports and handles all the applications and programs that a mobile device or computer uses. An OS employs a graphic user interface (GUI), a mixture of text and graphics that enables you to interact with the computer or device. Because of their integral nature, every smart device or computer requires an OS to perform its tasks and run its applications.

OS employs two components to handle computer applications and programs, the kernel and the shell. The kernel is the system's core inner component that processes the computer's data at the hardware level. It handles memory and process and input-output management. In contrast, the shell is the system's outer layer that handles the interaction between the OS and the user. The shell interacts with the OS by receiving input from a shell script or the user. Shell scripts are a sequence of system commands that the computer stores in a file.

【Ex.5】通过 Internet 查找资料，借助电子词典、辅助翻译软件及 AI 工具，完成以下技术报告，并附上收集资料的网址。通过 E-mail 发送给老师，或按照教学要求在网上课堂提交。
1. 当前世界上有哪些最主要的半导体生产厂家及有哪些最新型的产品（附上各种最新产品的图片）。
2. 当前世界上有哪些最主要的计算机芯片生产厂家及有哪些最新型的产品（附上各种最新产品的图片）。

Text B

Integrated Circuits

Integrated circuits, also known as microchips, play a crucial role in modern electronics. These tiny devices can contain hundreds of thousands of transistors, resistors, and other components, allowing them to perform complex tasks in a compact and efficient manner. There are many different types of integrated circuits available, each designed to meet specific needs and requirements. From memory chips used in computers to microcontrollers found in automobiles, integrated circuits come in a wide range of shapes, sizes, and specifications.

1. What Is an Integrated Circuit?

An integrated circuit can be defined as a circuit where all or some circuit elements are closely and inseparably connected with each other, where it acts as a single unit for multi-functional purposes.

An integrated circuit has multiple sets of electronic components on a single flat piece which is made up of semiconductor material. MOSFETs (Metal Oxide Semiconductor Field Effect Transistors) are integrated into a very small chip. An integrated circuit is faster, more reliable, more efficient and less expensive.

An integrated circuit generally comprises parts like transistors, resistors and condensers. It has multiple terminals for different types of functions to perform, especially embedded systems. These integrated circuits are miniature and can be formed on a silicon wafer.

2. Types of Integrated Circuit

2.1 Classification by chip size

(1) SSI (Small Scale Integration): SSI defines the initial phase of integrated circuit development. It features a limited number of components (3 to 30 gates) within a chip. It is used for simple circuit designs like basic logic gates, decoders, and multiplexers.

(2) MSI (Medium Scale Integration): an MSI chip has an internal capacity of 30 to 300 gates. It empowers circuits with capabilities such as arithmetic functions, data processing, and control systems. It is suitable for applications like subtractor, adder, and versatile register.

(3) LSI (Large Scale Integration): LSI represents a significant milestone in IC chip advancement. An LSI chip can host a complete subsystem (300 to 3000 gates) on a single chip. This enables easier manufacturing of microprocessor memory units and intricate digital functions, It contributes to various electronic projects including communication devices and computers.

(4) VLSI (Very Large Scale Integration): the pinnacle of IC technology, VLSI brings a revolution to the design and manufacturing of electronic systems. It allows configuring over three thousand gates on a single chip. Currently, the cutting-edge industry heavily relies on VLSI for advanced signal processors, microcontrollers, and application specific integrated circuits (ASICs).

(5) SLSI (Super Large Scale Integration): this group comes within the range of 10,000 to 100,000 transistors connected together. All these are connected on a single chip and collectively, and they perform the computational operations. Microcontroller and microprocessor chips are examples in this case.

(6) ULSI (Ultra Large Scale Integration): this system consists of more than millions of transistors working together. These are highly efficient and examples of such systems are CPU, GPU, video processor, etc.

2.2 Classification by chip thickness

(1) Thin film IC: manufactured by depositing a thin layer of resistive and conductive material on a substrate using techniques like sputtering or chemical vapor deposition (CVD), it offers higher precision. It is suitable for projects involving precise resistors and capacitors.

(2) Thick film IC: with thicker deposited layers, it's easier to construct, making it more cost-effective under similar conditions. It can handle high-power levels, making it best suited for projects like voltage regulators and amplifiers.

(3) Monolithic IC: it integrates different components like resistors, capacitors, transistors, and diodes onto a single semiconductor substrate made of silicon. Due to tightly interconnected components, it enhances performance, reliability, and reduces power consumption.

(4) Hybrid or multi-chip IC: utilizing wire bonding or flip-chip methods, it interconnects multiple chips. Designers can optimize individual component capabilities based on needs, significantly enhancing project customization and flexibility.

2.3 Classification by chip function

(1) Digital integrated circuit: considered the backbone of modern computing and communication systems, it processes binary data and manipulates signals with two possible values, 0 and 1. Examples include microprocessors, digital signal processors, and microcontrollers.

(2) Analog integrated circuit: it is used to process continuous signals that smoothly change over time. Examples are operational amplifiers, voltage regulators, and analog to digital converters (ADCs).

(3) Mixed signal integrated circuit: it is a combination of analog and digital components on a single board. It enables interaction between the digital and real-world domains. It requires precise control and applications involving analog-to-digital and digital-to-analog conversion.

(4) Power management IC: it regulates and distributes power within electronic systems, and ensures power efficiency while extending battery life. Examples include voltage regulators, power converters, and battery charging ICs.

(5) RF IC: it forms the core of wireless communication systems. It is equipped with oscillators, RF amplifiers, transceivers, and mixers to process high-frequency signals. It fosters the development of WiFi, smartphones, and satellite communication systems.

(6) Microcontroller: equipped with memory, a central processing unit, and input/output interfaces, it forms a complete computing system suitable for Internet of Things devices, embedded systems, and automation projects.

(7) Memory IC: it includes flash memory, random access memory (RAM), read-only memory

(ROM), and EEPROM. It provides storage and retrieval functions for digital information.

(8) Sensor IC: it converts real-world physical phenomena like temperature, light, pressure, and motion into electrical signals.

(9) Application specific integrated circuit (ASIC): it has custom-designed components for specific applications. It optimizes performance and efficiency by reducing unnecessary components. It is commonly used in cryptography, image processing, and signal processing operations.

(10) Field programmable gate array (FPGA): it is a general purpose IC used to execute certain operations after manufacturing. Examples include digital signal processing, PCB prototyping, and hardware acceleration.

(11) System on chip (SoC): it integrates various functions on a single chip, such as communication, memory, processing, and I/O interfaces.

(12) Voltage regulator module (VRM): it is essential for sensitive devices, VRM regulation provides voltage to electronic components to ensure effective and stable power supply.

(13) Clock generator: it produces precise timing signals to synchronize different components within electronic systems. It is a critical component for maintaining synchronization and data integrity.

(14) Display driver IC: it configures pixel data and refresh rates. It controls the display functionality of devices like monitors and mobile phones and ensures correct and smooth visual output.

(15) Audio amplifier: it is used to amplify and process audio signals. It is commonly applied in consumer electronics like headphones, audio players, and speakers.

3. Basic Features of Integrated Circuit

The following are the basic features of an integrated circuit:

(1) Construction: an integrated circuit is built on a silicon wafer, it has a miniature size, but is also made up of a conducting material.

(2) Packaging: a series of gold or aluminium wires are joined on the circuit board in the form of a printed circuit board, with metal tips on the outside to keep the components intact and connected to each other.

(3) Integrated circuit size: the size of an integrated circuit is generally between 1 square nm to around 200 square nm.

4. Advantages of Integrated Circuits

(1) More durable and reliable: they are typically more reliable, require less power, and are easier to manufacture. Additionally, they can be used to create complex products that would otherwise be difficult to achieve with discrete components.

(2) Cost effective: integrated circuits also offer a number of cost benefits. They are usually much less expensive than discrete components, and their small size allows them to be used in a wide range of applications. In addition, integrated circuits can be mass-produced quickly and

cheaply, making them ideal for large-scale production.

(3) More versatile: finally, integrated circuits are very versatile. They can be used to create a variety of different products, from basic logic gates to complex microprocessors. This makes them ideal for a wide range of applications, from consumer electronics to medical equipment.

5. Disadvantages of Integrated Circuits

(1) Increased repair costs: when a part of the circuits in the integrated circuit fails, it is usually necessary to replace the whole block, which increases the repair cost.

(2) Inconvenient fault diagnosis: relative to discrete electronic component circuits, it is not convenient to accurately determine the faults of integrated circuits when troubleshooting certain special faults.

(3) Difficulties in circuit disassembly: there are many pins on the integrated circuit, which brings great difficulties to repair and disassemble the integrated circuit. Especially for the four-column integrated circuit with many pins, the disassembly is difficult.

New Words

microchip	[ˈmaɪkrəʊtʃɪp]	n. 微芯片，微晶片；微型集成电路片
automobile	[ˈɔːtəməbiːl]	n. 汽车
shape	[ʃeɪp]	n. 形状；模型；状态
		vi. 使成形；形成
inseparably	[ɪnˈseprəblɪ]	adv. 不能分离地，不可分地
multi-functional	[mʌltɪˈfʌŋkʃənl]	adj. 多功能的
set	[set]	n. 集合；一套，一副
semiconductor	[ˌsemikənˈdʌktə]	n. 半导体
comprise	[kəmˈpraɪz]	vt. 包含，包括；由……组成，由……构成
condenser	[kənˈdensə]	n. 电容器
miniature	[ˈmɪnətʃə]	adj. 小型的，微小的
silicon	[ˈsɪlɪkən]	n. 硅
decoder	[ˌdiːˈkəʊdə]	n. 解码器
empower	[ɪmˈpaʊə]	vt. 准许；使能够
subtractor	[səbˈtræktə]	n. 减法器
adder	[ˈædə]	n. 加法器
versatile	[ˈvɜːsətaɪl]	adj. 多用途的；多功能的
register	[ˈredʒɪstə]	n. 寄存器
milestone	[ˈmaɪlstəʊn]	n. 里程碑，划时代事件
subsystem	[ˈsʌbˈsɪstəm]	n. 子系统，分系统
intricate	[ˈɪntrɪkət]	adj. 错综复杂的；难理解的
pinnacle	[ˈpɪnəkl]	n. 顶峰，顶点
collectively	[kəˈlektɪvlɪ]	adv. 全体地，共同地

sputtering	['spʌtərɪŋ]	n. 反应溅射法
substrate	['sʌbstreɪt]	n. 底层，基底
hybrid	['haɪbrɪd]	adj. 混合的
flip-chip	[flɪp tʃɪp]	n. 倒装法，倒装芯片
customization	['kʌstəmaɪzeɪʃən]	n. 用户化，专用化；定制
mixer	['mɪksə]	n. 混合器
phenomena	[fə'nɒmɪnə]	n. 现象
cryptography	[krɪp'tɒgrəfɪ]	n. 密码系统，密码术
acceleration	[əkˌselə'reɪʃn]	n. 加速
stable	['steɪbl]	adj. 稳定的
pixel	['pɪksl]	n. 像素
monitor	['mɒnɪtə]	n. 显示屏，屏幕；显示器；监测仪
audio	['ɔːdɪəʊ]	adj. 声音的；音频的
construction	[kən'strʌkʃn]	n. 结构
packaging	['pækɪdʒɪŋ]	n. 封装，包装
durable	['djʊərəbl]	adj. 耐用的，耐久的 n. 耐用品
versatile	['vɜːsətaɪl]	adj. 多用途的；多功能的
inconvenient	[ˌɪnkən'viːnɪənt]	adj. 不方便的，麻烦的
diagnosis	[ˌdaɪəg'nəʊsɪs]	n. 诊断，判断
disassembly	[ˌdɪsə'semblɪ]	n. 拆卸，分解

Phrases

tiny device	微型器件
a wide range of	大范围的，广泛的
be defined as	被定义为，被称为
single flat piece	单片
be made up of	由……组成，由……构成
embedded system	嵌入式系统
silicon wafer	硅片
initial phase	初期，初始阶段
logic gate	逻辑门
arithmetic function	算术函数
data processing	数据处理
be suitable for ...	适合……的
contribute to ...	对……有贡献，贡献于……
communication device	通信设备
cutting-edge industry	尖端行业
signal processor	信号处理器

video processor	视频处理器
thin film IC	薄膜集成电路
thick film IC	厚膜集成电路
deposited layer	沉积层
suit for	适合
voltage regulator	电压调节器，调压器，稳压器
monolithic IC	单片集成电路
power consumption	耗电量，能量消耗
multi-chip IC	多芯片集成电路
digital integrated circuit	数字集成电路
analog integrated circuit	模拟集成电路
continuous signal	连续信号
a combination of ...	……的组合
power converter	电源转换器
wireless communication system	无线通信系统
satellite communication system	卫星通信系统
image processing	图像处理
sensitive device	灵敏设备，敏感装置
power supply	电源
display driver	显示驱动器
refresh rate	刷新率
conducting material	导电材料
aluminium wire	铝丝，铝线
discrete component	分立元件
large-scale production	大规模生产
medical equipment	医疗器械
repair cost	修理成本，维修费用

Abbreviations

IC (Integrated Circuit)	集成电路
MOSFET (Metal Oxide Semiconductor Field Effect Transistor)	金属-氧化物-半导体场效应晶体管
SSI (Small Scale Integration)	小规模集成电路
MSI (Medium Scale Integration)	中规模集成电路
LSI (Large Scale Integration)	大规模集成电路
VLSI (Very Large Scale Integration)	甚大规模集成电路
ASIC (Application-Specific Integrated Circuit)	专用集成电路
SLSI (Super Large Scale Integration)	超大规模集成电路
ULSI (Ultra Large Scale Integration)	甚超大规模集成电路

CPU (Central Processing Unit)	中央处理器
GPU (Graphics Processing Unit)	图形处理器
CVD (Chemical Vapor Deposition)	化学气相沉积
ADC (Analog to Digital Converter)	模数转换器
RF (Radio Frequency)	射频
WiFi (Wireless Fidelity)	一种无线传输数据的技术
RAM (Random Access Memory)	随机存储器
ROM (Read-only Memory)	只读存储器
FPGA (Field Programmable Gate Array)	现场可编程门阵列
PCB (Printed-circuit Board)	印制电路板
VRM (Voltage Regulator Module)	电压调节模块
SoC (System on Chip)	单片系统
I/O (Input/Output)	输入/输出

【Ex.6】根据文章所提供的信息判断正误。

1. Integrated circuits can contain hundreds of thousands of transistors, resistors, and other components.
2. An integrated circuit has multiple sets of electronic components on a single flat piece which is made up of conductive material.
3. An MSI chip has an internal capacity of 300 to 3000 gates.
4. Currently, the cutting-edge industry heavily relies on VLSI for advanced signal processors, microcontrollers, and application specific integrated circuits (ASICs).
5. Analog integrated circuit is considered the backbone of modern computing and communication systems.
6. Memory IC includes flash memory, random access memory (RAM), read-only memory (ROM), and EEPROM. It provides storage and retrieval functions for digital information.
7. A series of gold or aluminium wires are joined on the circuit board in the form of a printed circuit board, with metal tips on the inside to keep the components intact and connected to each other.
8. The size of an integrated circuit is generally between 1 square nm to around 200 square nm.
9. Integrated circuits can be mass-produced quickly and cheaply, making them ideal for large-scale production.
10. When a part of the circuits in the integrated circuit fails, it is usually not necessary to replace the whole block.

科技英语翻译知识

词义的增减

英汉两种语言有不同的表达方式。在翻译过程中，要对语意进行必要的增减。在句子结构不完善，句子含义不明确或词汇概念不清晰时，需要对语意加以补足；反之，原文中的有些词如果在译文中不言而喻，就要省略一些不必要的词，使得译文更加严谨、明确。

1．增补

增补的基本功能是明确原文词汇含义，表达原文语法概念，满足译文修辞要求。

(1) The dielectric material in a practical capacitor is not perfect, and a small leakage current will flow through it.

实际应用的电容电介材料是不完全绝缘的，会有很小的漏电流流过。

not perfect 的意思是"不完全的、不完美的"，但在哪方面不完全，没有给出，因此要补足原文词汇含义"绝缘"。

(2) Electricity is convenient and efficient.

电用起来方便而有效。

增加"用起来"，明确原文含义。

(3) The high-altitude plane was and still is a remarkable bird.

高空飞机过去是而且现在还是一种了不起的飞行器。

增加"过去""现在"二词，表达动词过去时和现在时的语法概念。

(4) It is estimated that the new synergy between computers and net technology will have significant influence on the industry of the future.

有人预测，新的计算机和网络技术的结合将会对未来工业产生巨大的影响。

增加主语"有人"，使句子变成主动句。

(5) The solution of parallel AC circuits problems is different from series AC circuits.

并联交流电路问题的解答方法，不同于串联交流电路。

The solution of 的意思是"……的方法"，增加"解答"二字，更符合汉语表达习惯。

(6) This magnetic field may be that of a bar magnet, an U-shaped magnet, or an electromagnet.

这个磁场可以是条形磁铁的磁场，可以是马蹄形磁铁的磁场，也可以是电磁铁的磁场。

译文2次重复"可以是"和"磁场"，以强调有3种可能性。

(7) Now human beings have not yet progressed as to be able to make an element by combining protons, neutrons and electrons.

目前人类没有进展到能把质子、中子、电子三者化合成为一个元素的地步。

增加概括性数量词"三者"。

(8) Using a transformer, power at low voltage can be transformed into power at high voltage.

如果使用变压器,低电压的电力就能转换成高电压的电力。

增加"如果",使语气连贯。

(9) Were there no electric pressure in a conductor, the electron flow would not take place in it.

导体内如果没有电压,便不会产生电子流动现象。

增加"现象"一词,符合汉语习惯。

2. 省略

英语中的冠词、代词和连词在译文中往往可以省略。

(1) Hence the unit of electric current is the ampere.

因此,电流的单位是安培。

两个定冠词 the 省略。

(2) The diameter and the length of the wire are not the only factors to influence its resistance.

导线的直径和长度不是影响电阻的唯一因素。

省略物主代词 its。

(3) Like charges repel each other while opposite charges attract.

同性电荷相斥,异性电荷相吸。

省略连词 while。

(4) When atoms are joined together they form a larger particle called a molecule.

原子连接在一起就形成了更大的粒子,叫作分子。

省略连词 when。

(5) When we talk of electric current, we mean electrons in motion.

当我们谈到电流时,我们指的是运动的电子。

省略介词 in。

(6) Evidently semi-conductors have a lesser conducting capacity than metals.

显然,半导体的导电能力比金属差。

省略动词 have。

(7) There are many kinds of atoms, differing in both mass and properties.

原子种类很多,质量与性质都不相同。

省略引导词 there。

(8) The invention of radio has made it possible for mankind to communicate with each other over a long distance.

无线电的发明使人类有可能进行远距离通信。

省略形式宾语的引导词 it。

(9) Insulators in reality conduct electricity but, nevertheless, their resistance is very high.

绝缘体实际上也能导电,但其电阻很高。

省略重复性词语 nevertheless。

(10) Semi-conductors devices have no filament or heaters and therefore require no heating power or warmed up time.

半导体器件没有灯丝和加热器,因此不需要加热功率或加热时间。

省略重复性词语 and。

Reading Material

阅读下列文章。

Text	Note

Introduction to AutoCAD Electrical
 	
  Are you a beginner looking to learn AutoCAD Electrical? Look no further! In this passage, we will guide you through the fundamentals of AutoCAD Electrical and help you gain the skills necessary to navigate the software with confidence[1]. AutoCAD Electrical is a powerful tool used by electrical engineers and professionals to design and draft[2] electrical control systems. With its specialized features and functionalities, it streamlines the process of creating electrical drawings and automates various tasks. So, let's dive into this passage and unlock the world of AutoCAD Electrical!
 	
1. What is AutoCAD Electrical?
  AutoCAD Electrical is a software application developed by Autodesk that combines[3] the functionality of AutoCAD with additional tools specifically designed for electrical design tasks. It provides a comprehensive[4] set of features tailored to the needs of electrical engineers, allowing them to create, modify, and document electrical control systems efficiently.
 	
2. Why Learn AutoCAD Electrical?
  Learning AutoCAD Electrical can significantly enhance your productivity as an electrical engineer or professional. It offers numerous benefits, including:
  Efficiency: AutoCAD Electrical streamlines the design process and automates repetitive[5] tasks, reducing the time required to complete electrical drawings.
  Accuracy: with built-in electrical intelligence and automated
	[1] *n.* 信心,自信
	
	[2] *v.* 起草
	
	[3] *v.* 结合,合并,综合
	[4] *adj.* 全面的,综合性的
	
	[5] *adj.* 重复的

error-checking[6] features, AutoCAD Electrical helps minimize errors in designs, ensuring higher accuracy.

  Standardization: the software enables the use of standardized symbols, libraries, and templates[7], promoting consistency in electrical designs.

  Collaboration: AutoCAD Electrical supports seamless collaboration by allowing multiple users to work on the same project simultaneously, improving overall project efficiency.

  Integration: it seamlessly[8] integrates with other AutoCAD software and third-party applications, enabling smooth data exchange and enhancing workflow integration.

3. System Requirements

  Before you begin your journey with AutoCAD Electrical, ensure that your computer meets the minimum system requirements. The software demands a robust configuration to deliver optimal performance. The system requirements may vary depending on the version you are using, so it's always recommended to check the Autodesk website[9] for the latest information.

4. Exploring the Interface

4.1 The Ribbon

  The Ribbon is a central component of the AutoCAD Electrical interface. It provides access to various tools and commands. It is organized into tabs, each containing multiple panels[10]. It offers an intuitive and user-friendly way to navigate the software and access its extensive functionalities.

4.2 The Project Manager

  The Project Manager in AutoCAD Electrical serves as a hub for managing your electrical projects. It allows you to organize drawings, symbols, and other project-related files in a hierarchical[11] structure. The Project Manager simplifies project management and ensures easy access to all project components.

4.2.1 Creating and modifying drawings

  To create a new drawing in AutoCAD Electrical, follow these steps:

---

[6] *adj.* 验错的，检验误差的

[7] *n.* 样板，模板

[8] *adv.* 无缝地

[9] *n.* 网站

[10] *n.* 仪表板，面板

[11] *adj.* 等级的，分层的

Launch the software and open the Project Manager. Select the project in which you want to create the drawing.

Right-click on the "Drawings" folder[12] and choose "New Drawing" from the context menu.

Specify the drawing properties, such as the size and units, and click "OK".

AutoCAD Electrical provides a range of tools to modify your drawings quickly. Some common modification tasks include:

Moving and Copying Objects: use the "Move" and "Copy" commands to relocate[13] or duplicate objects within your drawing.

Editing Objects: modify the properties[14] of objects using commands like "Properties" or "Properties Palette."

Adding Components: to add components, select the appropriate symbol from the Symbol Library and place it in your drawing. You can browse the library by clicking on the "Schematic" tab in the Ribbon and selecting "Symbol Builder" or by using the "Insert Component" command. Once the symbol is inserted[15], you can edit its properties and connect it to other components.

Removing Components: to remove components, select the component you want to delete and press the "Delete" key on your keyboard. You can also use the "Erase" command from the Ribbon. Be cautious when removing components, as it may affect the overall functionality of your electrical design.

4.2.2　Creating and editing circuits

AutoCAD Electrical provides a seamless way to create circuits in your electrical designs. To create a circuit, follow these steps:

Insert Components: start by inserting the necessary components into your drawing. You can use the Symbol Library to select and place the required symbols.

Connecting Components: after placing the components, use the "Wire" command to connect[16] them. Click on the first component's connection point and then click on the second component's connection point to establish the connection. AutoCAD Electrical automatically assigns wire numbers and maintains the connectivity between components.

Modifying Circuits: you can easily modify circuits by adding or removing components, repositioning[17] wires, or changing wire properties. AutoCAD Electrical's intelligent features help maintain the circuit integrity throughout the modifications[18].

[12] *n.* 文件夹

[13] *v.* 迁移，重新安置
[14] *n.* 属性，特性

[15] *vt.* 插入，嵌入

[16] *v.* 连接，接通

[17] *vt.* 复位；改变……的位置 *n.* 重新配置，复位
[18] *n.* 修改，变更

To edit circuits in AutoCAD Electrical, follow these steps:

Select the Circuit: click on the circuit you want to edit to select it. You can use the "Select" command or simply click on the components and wires that make up the circuit.

Modify the Circuit: once the circuit is selected, you can make changes to its properties, add or remove components, or modify the wire connections. AutoCAD Electrical provides a range of editing tools, such as "Edit Component" and "Edit Wire", to help you make precise modifications.

4.2.3　Generating reports and documentation

AutoCAD Electrical offers powerful reporting and documentation capabilities which allow you to generate comprehensive reports for your electrical designs. These reports provide valuable information about the components, wires, and other aspects of your electrical systems. Here are some commonly used reports in AutoCAD Electrical:

Bill of Materials (BOM): the BOM report provides a list of all the components used in your electrical design, along with their quantities and other relevant details[19]. It helps in procurement[20] and project costing.

Wire Connection Report: this report displays the connections between wires and components in your electrical design. It helps in troubleshooting and maintenance.

Terminal Strip Report: the terminal strip report shows the arrangement of wires on terminal strips. It assists in panel wiring and installation.

To generate reports in AutoCAD Electrical, follow these steps:

Open the Project Manager: launch the Project Manager and navigate to the project that contains the drawing for which you want to generate a report.

Select the Drawing: right-click on the drawing and choose "Reports" from the context menu. Select the desired report from the list.

Configure Report Settings: configure the settings for the report, such as the content, formatting, and output format. You can customize the report according to your requirements.

Generate[21] the Report: click on the "Generate" button to create the report. AutoCAD Electrical will generate the report based on the

[19] *n.* 细节，详情
[20] *n.* 采购
[21] *vt.* 形成，造成，产生

| selected settings and display it on the screen. You can save the report in various formats, such as PDF or Excel, for further use. | |

参考译文

微控制器

  微控制器是一种紧凑的集成电路，旨在管理嵌入式系统中的特定操作。典型的微控制器包括单个芯片上的处理器、存储器和输入/输出（I/O）外设。

  微控制器有时被称为嵌入式控制器或微控制器单元（MCU），广泛应用于车辆、机器人、办公机器、医疗器械、移动无线电收发器、自动售货机和家用电器等设备中。它们本质上是简单的微型个人计算机（PC），旨在控制较大部件的小功能，它们没有复杂的前端操作系统（OS）。

1. 微控制器如何工作？

  微控制器嵌入系统内部以控制设备中的特定功能。它通过使用中央处理器解释从 I/O 外设接收的数据来实现这一点。微控制器接收到的临时信息存储在其数据存储器中，处理器访问该信息并使用存储在程序存储器中的指令来解读和应用输入的数据。然后，它使用其 I/O 外围设备进行通信并执行适当的操作。

  微控制器被广泛用于各种系统和设备。设备通常利用其中的多个微控制器协同工作来处理各自的任务。

  例如，一辆汽车可能有很多微控制器，用于控制内部的各种独立系统，如防抱死制动系统、牵引力控制、燃油喷射或悬架控制。所有微控制器相互通信以告知正确的操作。有些可能与汽车内更复杂的中央计算机通信，而另一些可能仅与其他微控制器通信。它们使用 I/O 外设发送和接收数据，并处理该数据以执行指定的任务。

2. 微控制器由哪些元件组成？

2.1 处理器（CPU）

  处理器可以被视为该设备的大脑。它处理并响应指导微控制器功能的各种指令。这涉及执行基本算术运算、逻辑运算和 I/O 操作。它还执行数据传输操作，并将命令传递给大型嵌入式系统中的其他组件。

2.2 存储器

  微控制器的存储器用于存储处理器接收并使用的数据，来响应要执行的编程指令。微控制器的存储器有以下两种主要类型。

（1）程序存储器，它存储有关 CPU 执行的指令的长期信息。程序存储器是非易失性存储器，这意味着它可以在不需要电源的情况下长期保存信息。

（2）数据存储器，它是在执行指令时用于临时数据存储所需的存储器。数据存储器是易失性的，这意味着它保存的数据是临时的，并且只有在设备连接到电源时数据才会被存储。

2.3　I/O 外设

输入和输出设备是处理器与外界的接口。输入端口接收信息并将其以二进制数据的形式发送到处理器。处理器接收该数据并将必要的指令发送到执行微控制器外部任务的输出设备。

虽然处理器、存储器和 I/O 外设是微处理器的典型元件，但常常还包含其他元件。I/O 外设一词仅指与存储器和处理器连接的支持组件。有许多支持组件可以归类为外围设备。

2.4　模数转换器（ADC）

ADC 是将模拟信号转换为数字信号的电路。它让微控制器中心的处理器与外部模拟设备（如传感器）连接。

2.5　数模转换器（DAC）

DAC 的功能与 ADC 相反，并让微控制器中心的处理器将其输出信号传送到外部模拟部件。

2.6　系统总线

系统总线是将微控制器的所有组件连接在一起的连接线。

2.7　串行端口

串行端口是 I/O 端口的一个示例，它把微控制器连接到外部部件。它具有与 USB 或并行端口类似的功能，但交换信息的方式有所不同。

3. 微控制器的特点

微控制器的处理器会因应用而异。选项范围从简单的 4 位处理器、8 位处理器或 16 位处理器到更复杂的 32 位处理器或 64 位处理器。微控制器可以使用易失性存储器和非易失性存储器。易失性存储器类型包括随机存储器（RAM），非易失性存储器类型包括闪存、可擦编程只读存储器（EPROM）和电擦除可编程只读存储器（EEPROM）。

一般来说，微控制器被设计为易于使用，无须额外的计算部件，因为它们被设计有足够的板载内存，并提供用于一般 I/O 操作的引脚，所以它们可以直接与传感器和其他部件连接。

微控制器架构可以基于哈佛架构或冯·诺依曼架构，两者提供了在处理器和存储器之间交换数据的不同方法。在哈佛架构中，数据总线和指令是分开的，允许同时传输。对于冯·诺依曼架构，数据和指令用的是同一条总线。

微控制器处理器可以基于复杂指令集计算（CISC）或精简指令集计算（RISC）。虽然CISC更容易实现并且内存使用效率更高，但由于执行指令所需的时钟周期数较多，它可能会导致性能下降。RISC处理器简化了指令集，从而提高了设计的简单性。

微控制器首次面世时，它们仅使用汇编语言。如今，C编程语言是一种流行的选择。其他常见的微处理器语言包括Python和JavaScript。

MCU具有输入和输出引脚来实现外设功能。这些功能包括模数转换器、液晶显示（LCD）控制器、实时时钟（RTC）、通用同步/异步收发器（USART）、定时器、通用异步收发器（UART）及通用串行总线（USB）的连接。

4. 常用微控制器

（1）Intel MCS-51，通常称为8051微控制器，于1985年首次开发。
（2）Atmel于1996年开发的AVR微控制器。
（3）Microchip Technology的可编程接口控制器（PIC）。
（4）各种获得许可的高级精简指令集计算机（ARM）微控制器。

5. 微控制器的应用

微控制器用于多个行业和应用领域，包括家庭和企业、楼宇自动化、制造业、机器人、汽车、照明、智能能源、工业自动化、通信和物联网（IoT）部署。

微控制器的一个非常具体的应用是用作数字信号处理器。输入的模拟信号常常带有一定程度的噪声。在这种情况下，噪声意味着不明确的值，它们无法轻易转换为标准数字值。微控制器可以使用其ADC和DAC将输入的有噪声的模拟信号转换为干净的输出数字信号。

最简单的微控制器可促进日常便利用品中机电系统的操作，如烤箱、冰箱、烤面包机、移动设备、遥控钥匙、视频游戏系统、电视和草坪浇水系统。它们在复印机、扫描仪、传真机和打印机等办公设备及智能电表、ATM和安全系统中也很常见。

更复杂的微控制器在飞机、航天器、远洋船舶、车辆、医疗和生命支持系统及机器人中执行关键功能。

6. 影响微控制器选择的因素

为特定应用选择微控制器时，需要考虑以下几个因素。

（1）处理能力：任务的复杂性决定了微控制器所需的处理能力，这会影响在8位微控制器、16位微控制器或32位微控制器之间的选择。
（2）内存容量：足够的内存对于存储程序代码、数据和变量至关重要。具有较大代码库或数据存储要求的应用程序需要更多内存。
（3）外设支持：根据应用，必须考虑所需的外设，如ADC、DAC、定时器和通信接口。
（4）电源效率：电池供电的设备需要选择优化后低功耗的微控制器，以延长电池寿命。
（5）成本：项目预算额度可能会影响微控制器的选择，因为高性能的微控制器往往更昂贵。

Text A

Analog Circuits and Digital Circuits

Both analog circuits and digital circuits are extensively used in various fields of electrical and electronic engineering for signal processing.

1. What Is an Analog Circuit?

An analog circuit is a type of electronic circuit that can process any analog signal and produce an output in analog form (see Figure 5-1). Analog circuits are composed of resistors, inductors and capacitors, etc.

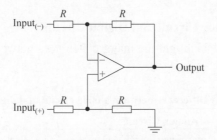

Figure 5-1　Simple Analog Circuit

The type of signal which is a continuous function of time is known as an analog signal. All the real-world signals are the analog signals, therefore, the analog circuit do not require any conversion of the input signal i.e. the analog input signal can be directly fed to the analog circuit without any loss and it can be directly processed by the given analog circuit. Also, the output signal produced by the analog circuit is an analog signal.

Based on the circuit behavior and the components used, the analog circuit can be of two types. They are active circuit and passive circuit. Amplifiers are the examples of active analog circuit while low pass filter is an example of passive circuit. The main drawback of the analog circuits is that the analog signals are very susceptible to noises, and the noises may cause distortion of the signal waveform and the loss of information.

2. What Is a Digital Circuit?

A digital circuit is an electronic circuit that processes digital signals. A signal that is a discrete function of time is known as a digital signal.

The basic building blocks of digital circuits are digital logic gates (see Figure 5-2). The digital circuit can process only digital signals, but the real-world signals are of analog nature. Therefore, they need to be converted into digital signals using special electronic circuit known as analog to digital converter (ADC). The output of the digital circuits is also digital signals, which is required to be converted back into the analog signal.

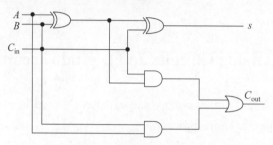

Figure 5-2 Simple Digital Circuit

There may be loss of information in the digital circuit during sampling process. The digital circuits can only be active circuits, which means they require an additional power source to power the circuit. [1]

3. Difference Between Analog Circuit and Digital Circuit

The following Table 5-1 highlight the major differences between analog circuits and digital circuits.

Table 5-1 Difference between analog circuit and digital circuit

Parameter	Analog Circuit	Digital Circuit
Definition	The electronic circuit which can process only analog signals is known as analog circuit.	The electronic circuit which can process only digital signals is known as digital circuit.
Input signal	The input signal to the analog circuit must be a continuous time signal or analog signal.	The input signal to the digital circuit is a discrete time signals or digital signal.
Output signal	Analog circuits produces output in the form of analog signals.	The output of the digital circuit is a digital signal.
Circuit components	The circuit components of the analog circuits are resistors, inductors, capacitors, etc.	The main circuit components of the digital circuits are logic gates.
Need of converters	The analog circuits can process the analog signals which exist in the nature directly. Therefore, analog circuits do not require signal converters.	The digital circuits can process signals only in digital form. Thus, digital circuits require signal converter, i.e. analog to digital converter (ADC) and digital to analog converter (DAC).
Susceptibility to noise	The analog signals are more susceptible to noises.	The digital signals are immune to noises.

续表

Parameter	Analog Circuit	Digital Circuit
Design	The analog circuits are complex to design because their circuit components need to be placed manually.	The designing of complicated digital circuits is relatively easier by using multiple software.
Flexibility	The implementation of analog circuit is not flexible.	The digital circuits offer more flexible implementation process.
Types	Analog circuits can be of two types: active circuit and passive circuit.	Digital circuits are of only one type named active circuit.
Processing speed	The processing speed of analog circuits is relatively low.	The digital circuits have higher processing speed than analog circuits.
Power consumption	The analog circuits consume more power.	The power consumed by the digital circuits is relatively less.
Accuracy & precision	The analog circuits are less accurate and precise.	The digital circuits are comparatively more accurate and precise.
Observational errors	In case of analog circuits, there may be an observational error in the output.	The digital circuits are free from observational errors in the output.
Signal transmission	In case of analog circuits, the signals are transmitted in the form of waves either wirelessly or with wires.	In the digital circuits, the signals can only be transmitted through wires in the digital form.
Form of information storage	The analog circuits store the information in the form of waves.	Digital circuits store the information in binary form.
Logical operations	The analog circuit are not able to perform the logical operations efficiently.	Digital circuit performs logical operations efficiently.

4. Analog to Digital Converters

Analog to digital converters, or ADCs, are essential components in modern electronic devices. They are used to convert analog signals, such as sound or light, into digital data that can be processed by computers and other digital devices. ADCs are used in a wide range of applications, from audio recording and playback to medical imaging and industrial control systems.

ADCs work by converting analog signals into a series of digital values (see Figure 5-3). The process of converting analog signals into digital data is a three-step process, consisting of sampling, quantization, and encoding.

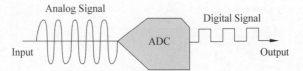

Figure 5-3 Analog to Digital Converter

4.1 Sampling

The first step in the conversion process is sampling. Sampling involves taking periodic samples of the analog signal at a fixed rate. The rate of sampling is known as the sampling rate, and it is measured in samples per second. The higher the sampling rate, the more samples are taken, and the more accurate the digital representation of the analog signal.

Sampling is a critical step in the conversion process because it determines the accuracy of the digital representation of the analog signal. If the sampling rate is too low, the digital representation of the analog signal will be inaccurate, and important details of the signal may be lost.

4.2 Quantization

The second step in the conversion process is quantization. Quantization involves assigning a digital value to each sample of the analog signal. The process of quantization involves dividing the input voltage range into a finite number of discrete steps. Each step is assigned a digital value, which represents the voltage of the input signal at that point in time.[2] The more steps in the quantization process, the more accurate the digital representation of the analog signal.

Quantization ensures that the digital representation of the analog signal is accurate and precise. Without quantization, the digital representation of the analog signal would be continuous, and it would be impossible to store or process the signal using digital devices.

4.3 Encoding

The final step in the conversion process is encoding. Encoding involves converting the digital values produced by the quantization process into binary codes that can be stored and processed by computers and other digital devices.[3] The most common encoding method is the binary code, where each digital value is represented by a string of 1's and 0's.

Encoding allows the digital representation of the analog signal to be stored and processed by computers and other digital devices. Without encoding, the digital representation of the analog signal would be meaningless to digital devices, and it would be impossible to use the signal for any practical purpose.

5. Digital to Analog Converter (DAC)

A digital to analog converter (DAC) is a device that converts digital signals into analog signals (see Figure 5-4). It is commonly used in various electronic systems, such as audio players, digital instruments, and communication devices. The process of converting a digital signal into an analog signal involves several steps.

Figure 5-4　Digital to Analog Converter

5.1 Input digital signal

The DAC receives a digital signal as input. This digital signal is usually in the form of binary data, where each bit represents a discrete value.

5.2 Sample-and-hold

The digital signal is first passed through a sample-and-hold circuit. This circuit samples the digital signal at regular intervals and holds each sample value until the next sample is taken. [4] This process helps to reconstruct the continuous-time analog signal from the discrete digital samples.

5.3 Conversion

The sampled digital values are then fed into a digital-to-analog conversion circuit. The purpose of the conversion circuit is to convert the discrete digital values into continuous analog voltages or currents. There are different types of DAC architectures, such as resistor ladder DACs, delta-sigma DACs, and multiplying DACs, each with its own implementation details. However, the basic principle involves using the digital input to generate a corresponding analog output.

5.4 Reconstruction filter

The output of the DAC, which is an analog signal, may contain some unwanted high-frequency components due to the sampling process. To remove these unwanted components, a reconstruction filter is often used. The reconstruction filter attenuates the high-frequency components and reconstructs a smooth analog signal.

5.5 Output amplification

In some cases, the output signal from the DAC may need to be amplified to achieve the desired signal level. An amplifier is used to increase the power or voltage level of the analog signal before it is sent to the output device, such as a speaker or a display.

New Words

analog	['ænəlɒːg]	n. 模拟
		adj. 模拟的
digital	['dɪdʒɪtl]	n. 数字
		adj. 数字的
form	[fɔːm]	n. 形式
continuous	[kən'tɪnjuəs]	adj. 连续的
function	['fʌŋkʃn]	n. 函数；功能，作用
conversion	[kən'vɜːʃn]	n. 转换，变换
behavior	[bɪ'heɪvjə]	n. 行为；（机器等的）运转状态
amplifier	['æmplɪfaɪə]	n. 放大器；扩音器

filter	['fɪltə]	n. 滤波器
		v. 过滤，渗透
susceptible	[sə'septəbl]	adj. 易受影响的
noise	[nɔɪz]	n. 噪声，杂音
definition	[ˌdefɪ'nɪʃn]	n. 定义
susceptibility	[səˌseptə'bɪləti]	n. 敏感性
complex	['kɒmpleks]	adj. 复杂的；复合的
manually	['mænjʊəlɪ]	adv. 用手地，手动地
flexibility	[ˌfleksə'bɪləti]	n. 柔韧性，机动性，灵活性
implementation	[ˌɪmplɪmen'teɪʃn]	n. 实现
consumption	[kən'sʌmpʃn]	n. 消耗，消费
observational	[ˌɒbzə'veɪʃənl]	adj. 观测的，观察的
wirelessly	['waɪələsly]	adv. 无电线地
storage	['stɔːrɪdʒ]	n. 贮存，储存
quantization	[ˌkwɒntɪ'zeɪʃən]	n. 量化，数字化
encode	[ɪn'kəʊd]	vt. 编码
periodic	[ˌpɪərɪ'ɒdɪk]	adj. 周期的，定期的
representation	[ˌreprɪzen'teɪʃn]	n. 表示，代表，表现
inaccurate	[ɪn'ækjərət]	adj. 不精确的，不准确的
meaningless	['miːnɪŋləs]	adj. 无意义的，无价值的
binary	['baɪnərɪ]	adj. 二进制的
bit	[bɪt]	n. 位，比特（二进位制信息单位）
interval	['ɪntəvl]	n. 间隔，区间
reconstruct	[ˌriːkən'strʌkt]	vt. 重建，改造
high-frequency	['haɪ'friːkwənsɪ]	adj. 高频率的
smooth	[smuːð]	adj. 平滑的，光滑的
		vt. 使平滑
achieve	[ə'tʃiːv]	v. 实现

Phrases

analog circuit	模拟电路
digital circuit	数字电路
signal process	信号处理
analog signal	模拟信号
be composed of…	由……组成
be fed to	馈送到
active circuit	有源电路
passive circuit	无源电路
signal waveform	信号波形

building block	组成模块
digital logic gate	数字逻辑门
be converted into	被转换为
sampling process	采样过程，抽样过程
power source	电源
input signal	输入信号
output signal	输出信号
circuit component	电路元件
processing speed	处理速度
logical operation	逻辑运算
industrial control system	工业控制系统
sampling rate	采样率，抽样率
digital value	数字值
digital device	数字设备，数字器件
binary code	二进制代码
encoding method	编码方法
a string of	一串，一系列
audio player	音频播放器
digital instrument	数字仪器
sample-and-hold circuit	采样保持电路
reconstruction filter	重建滤波器，重构滤波器

Abbreviations

ADC (Analog to Digital Converter)	模数转换器
DAC (Digital to Analog Converter)	数模转换器

Notes

[1] The digital circuits can only be active circuits, which means they require an additional power source to power the circuit.

本句中，which means they require an additional power source to power the circuit 是一个非限定性定语从句，对主句进行补充说明。

本句意为：数字电路只能是有源电路，这意味着它们需要额外的电源来为电路供电。

[2] Each step is assigned a digital value, which represents the voltage of the input signal at that point in time.

本句中，which represents the voltage of the input signal at that point in time 是一个非限定性定语从句，对 a digital value 进行补充说明。

本句意为：每个步骤都分配一个数字值，它代表该时间点输入信号的电压。

[3] Encoding involves converting the digital values produced by the quantization process into binary codes that can be stored and processed by computers and other digital devices.

本句中，produced by the quantization process 是一个过去分词短语，作定语，修饰和限定 the digital values。that can be stored and processed by computers and other digital devices 是一个定语从句，修饰和限定 binary codes。

本句意为：编码指将量化过程产生的数字值转换为可以由计算机和其他数字设备存储和处理的二进制代码。

[4] This circuit samples the digital signal at regular intervals and holds each sample value until the next sample is taken.

本句中，This circuit 是主语，samples the digital signal at regular intervals and holds each sample value 是 and 连接的并列谓语。until the next sample is taken 是时间状语。samples 是动词 sample 的单数第三人称形式，后面的两个 sample 是名词。

本句意为：该电路定期对数字信号进行采样，并保存每个采样值，直到进行下一个采样。

Exercises

【Ex.1】根据课文内容，回答以下问题。

1. What is an analog circuit? What are analog circuits composed of ?

2. What is a digital circuit?

3. What are the types of analog circuits and digital circuits respectively?

4. How do ADCs work? What are the three steps of doing it mentioned in the passage?

5. How many steps does the process of converting a digital signal into an analog signal involve? What are they?

【Ex.2】根据下面的英文解释，写出相应的英文词汇。

英 文 解 释	词 汇
a type of electronic circuit that uses continuous signals (analog signals) to represent information	
a type of electronic circuit that use discrete signals (digital signals) to represent information	
a system that converts a digital signal into an analog signal	
an electronic device that increases the voltage, current, or power of a signal	

续表

英 文 解 释	词 汇
a device or process that removes some unwanted components or features from a signal	
communication using radio waves, light, or other methods without wires	
a process through which digital data is saved within a data storage device by means of computing technology	
the process of converting data into a format required for a number of information processing needs	
a numeric system which uses two numerals, 0 and 1, to represent all real numbers	
the smallest unit of data that a computer can process and store	

【Ex.3】把下列句子翻译成中文。

1. The signal will be converted into digital code.

2. No computer can imitate the complex functions of the human brain.

3. Amplifier array provides a new perspective for low power consumption design.

4. Filters do not remove all contaminants from water.

5. Software business process needs higher flexibility and adaptability.

6. Any device that's built to receive a wireless signal at a specific frequency can be overwhelmed by a stronger signal coming in on the same frequency.

7. His task is to ensure the fair use and storage of personal information held on computer.

8. We compared the human mind to a computer which actively seeks information to process, encodes it and stores it for future use.

9. The instructions are translated into binary code, a form that computers can easily handle.

10. To give a definition of a word is more difficult than to give an illustration of its use.

【Ex.4】把下列短文翻译成中文。

Industrial control system (ICS) is a collective term used to describe different types of control systems and associated instrumentation, which include the devices, systems, networks, and controls used to operate and/or automate industrial processes. Depending on the industry, each ICS functions differently and are built to electronically manage tasks efficiently. Today the devices and protocols used in an ICS are used in nearly every industrial sector and critical infrastructure such as the manufacturing, transportation, energy, and water treatment industries. There are several types of ICSs, the most common of which are supervisory control and data acquisition (SCADA) systems, and distributed control systems (DCS).

【Ex.5】通过 Internet 查找资料，借助电子词典、辅助翻译软件及 AI 工具，给出模拟电路和数字电路的若干应用实例，并附上收集资料的网址。通过 E-mail 发送给老师，或按照教学要求在网上课堂提交。

Text B

Digital Circuit Elements

1. CMOS Element and Watch Switching

The complementary MOSFET scheme (or CMOS) started the second revolution in computational machines. The limits of speed and density were conquered by the move to semiconductors and very large scale integration, but the power consumption and circuit cooling demands of bipolar transistors packed at extreme densities were formidable problems. The problem was that the transistors were always "ON" (in other words drawing current and dissipating energy). CMOS circumvents this problem and allows bits to be stored without constant power consumption. A schematic of the CMOS inverter is given in the figure below (see Figure 5-5). The device dissipates energy only when it is switched from high to low or back. Quiescent operation in either the high or the low state dissipates essentially no power. So cooling the circuit is much easier, and supplying power is much less of a problem. If you don't believe me, just ask your calculator, digital watch or your laptop.

Connect V_{DD} = +5V and ground to the CD4007 pins as depicted below using only one set of transistors. For example, pin 10 = V_I, pin 11 = V_{DD}, pin 9 = GND and pin 12 = V_O. Connect a 500Ω resistor between V_{DD} and pin 11 for better performance.

Slowly ramp the input voltage from zero up to 3.5V. At some point the output should switch from high to low. Note the voltage where the switch occurs.

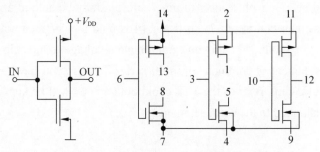

Figure 5-5　A schematic of the CMOS inverter

Now connect a 100Ω resistor in series with pin 9 above the ground point.

Try to measure the transient current (momentary voltage across the resistor) as you slowly ramp the input voltage up and down to make the output switch. If you can't see the signal you can cheat by using the Miller effect by adding a medium sized capacitor between output and ground.

Try to measure the intrinsic switching time and estimate the power consumption for such an inverter switched at 1MHz compared with a bipolar circuit where the devices are constantly passing current.

Estimate the power consumption for switching at 1GHz.

2. Gates, Truth Tables, and Pull-up Resistors

One of the simplest gates is the inverter. The Boolean equation for the inverter is

$$Y = \overline{A}$$

The following is the diagrammatic representation of the inverter (see Figure 5-6).

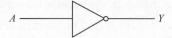

Figure 5-6　The diagrammatic representation of the inverter

The 7404 chip contains 6 inverters and can be schematically represented as follows (see Figure 5-7).

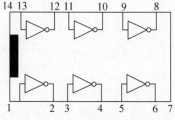

Figure 5-7　The 7404 chip

The diagram of the chip is drawn as if you are viewing it from above. Note the thick black line that is used as an orientation mark, located at the left end of the chip. Also note that pin 14 and pin 7 are not connected to an inverter, they are the power supply connections for the chip. Pin 14 must be held at +5V with respect to pin 7. No pin may be held at a voltage greater than that at pin 14 or less than that at pin 7. So if pin 14 is held at +5V then nothing can be greater than +5V and if pin 7 is grounded then you cannot have a voltage to the chip that is less than 0V. This supply pin assignment is common for 7400 series TTL. If you have a question about the wiring of a particular chip, refer to the TTL cookbook; a copy will be kept in the lab. You may also use the websites linked to the course homepage.

3. Procedure

(1) Wire a 7404 inverter, on the Digit-Designer as follows (see Figure 5-8).

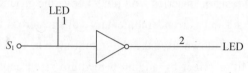

Figure 5-8　The 7404 inverter

(2) Apply a clock signal from the clock on the Digit-Designer Box to the input of one of the inverters in the 7404 IC chip. Simultaneously look at the input and output on the oscilloscope and also the LED's on the Digit-Designer for a range of clock frequencies from 1kHz to 100kHz.

Comment on the input and output of the circuit. Are there any timing problems with this circuit?

(3) Wire 6 inverters from the 7404 IC chip in series. Connect the Digit-Designer clock to the first inverter.

Observe the input of the 1st and the output of the last inverter simultaneously on the oscilloscope. Determine the "propagation delay" through a single inverter.

(4) Wire the following circuit using an open-collector inverter (7404) (see Figure 5-9). The 2kΩ "pull-up" resistor is necessary for speed and noise immunity when driving a TTL input. The numbers that you see on the diagram are to distinguish the input pins on the 7404 IC chip. So the 1 means that the input A should be connected to pin 1 on the 7404 IC chip and so on. The equation at the bottom right is the algebraic representation of the logical NOR equation.

Verify that the circuit performs the logical NOR function.

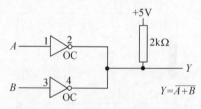

Figure 5-9　An open-collector inverter

4. B-2 Translating Boolean Equations into Electronic Circuits

In many cases, translation of Boolean equations into electronic logic can be accomplished by a straightforward, one-for-one replacement of a term or group of terms in the equation by a gate. As an illustration, we will consider the 4-input data selector or "multiplexer".

We have 4 digital signals that we would like to be able to send over a single wire. We need logic that defines the output Y of a circuit to be the nth input signal X_n, where the number, n, is given. Since n can take on 4 values, it must be a 2-digit binary number, which we will call BA. In other words, BA selects the input X_n, which will appear on the output. The truth table expressing this circuit is:

B	A	Y(output)
0	0	X_0
0	1	X_1
1	0	X_2
1	1	X_3

From this table, a Boolean equation can be written by inspection.

$$Y = (B \cap A \cap X_0) \cup (B \cap A \cap X_1) \cup (B \cap A \cap X_2) \cup (B \cap A \cap X_3)$$

From the Boolean equation above, the following schematic that describes the equation can be drawn (see Figure 5-10).

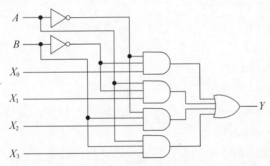

Figure 5-10　The schematic that describes the equation

In implementing the logic displayed in this diagram, we are slightly hampered by the fact that a 7400 series 4-input OR gate does not exist. The elegant solution to this problem involves DeMorgan's theorem and some common sense.

Verify that your circuit is working properly for 5 of the combinations in your truth table.

New Words

complementary	[ˌkɒmplɪˈmentrɪ]	*adj.* 补充的，补足的
scheme	[skiːm]	*n.* 安排，配置；计划，方案；图解，摘要
		vt. & vi. 计划，设计，图谋，策划
computational	[ˌkɒmpjuˈteɪʃənl]	*adj.* 计算的
density	[ˈdensətɪ]	*n.* 密度；浓度
integration	[ˌɪntɪˈgreɪʃn]	*n.* 综合，集成
bipolar	[ˌbaɪˈpəʊlə]	*adj.* 有两极的，双极的

formidable	[fə'mɪdəbl]	adj. 强大的，令人敬畏的，艰难的
circumvent	[ˌsɜːkəm'vent]	vt. 围绕，包围，智取
inverter	[ɪn'vɜːtə]	n. 逆变器，变极器；变流器；反相器
quiescent	[kwɪ'esnt]	adj. 静止的
laptop	['læptɒp]	n. 便携式计算机，膝上型计算机
pin	[pɪn]	n. 针；钉，销，栓
		vt. 钉住，别住，阻止，扣牢，止住，牵制
depict	[dɪ'pɪkt]	vt. 描述，描写
ramp	[ræmp]	n. 斜坡，坡道
		vi. 蔓延
		vt. 使有斜面
momentary	['məʊməntrɪ]	adj. 瞬间的，刹那间的
Boolean	['buːlɪən]	n. 布尔
		adj. 布尔的
gate	[geɪt]	n. 逻辑门
diagrammatic	[ˌdaɪəgrə'mætɪk]	adj. 图表的，概略的
orientation	[ˌɔːrɪən'teɪʃn]	n. 方向，方位，定位
chip	[tʃɪp]	n. 芯片，碎片
		vt. & vi. 削成碎片，碎裂
website	['websaɪt]	n. 网站
immunity	[ɪ'mjuːnətɪ]	n. 免疫性
translation	[trænz'leɪʃn]	n. 转化，转换；翻译，译文
multiplexer	['mʌltɪˌpleksə]	n. 多路（复用）器
theorem	['θɪərəm]	n. 定理，法则

Phrases

computational machine	计算机器
formidable problem	棘手的问题
dissipating energy	耗散能量，消散能量
transient current	瞬态电流，暂态电流，瞬变电流
Miller effect	米勒效应
very large scale integration	超大规模集成电路
inverter switch	逆变器开关；转换开关
truth table	真值表
pull-up resistor	上拉电阻，牵引电阻
Boolean equation	布尔方程
clock signal	时钟信号
clock frequency	时钟频率
propagation delay	传输延迟，传播延迟

DeMorgan's theorem 德莫根定理
common sense 常识

Abbreviations

CMOS (Complementary Metal Oxide Semiconductor) 互补金属氧化物半导体
MOSFET (Metallic Oxide Semiconductor Field Effect Transistor) 金属-氧化物-半导体场效应晶体管
TTL (Transistor-Transistor Logic) 晶体管-晶体管逻辑
LED (Lighting Emitting Diode) 发光二极管

Exercises

【Ex.6】根据文章所提供的信息判断正误。

1. The CMOS started the second revolution in computational machines.
2. Now, the limits of speed and density are formidable problems.
3. CMOS can store bit without constant power consumption.
4. Quiescent operation in either high or low state dissipates a little energy.
5. The 7404 chip contains 6 inverters.
6. None of the pins of the 7404 chip is connected to an inverter.
7. Pin 14 and pin 7 from the 7404 chip are the power supply connections for the chip.
8. For the 7404 chip, every pin may be held at a voltage greater than pin 7.
9. If pin 14 is held at a voltage +24V and pin 7 is held at a voltage +0V, the 7404 chip will work well.
10. For the 7404 IC chip, pin 14 must be held at +5V with respect to pin 7. No pin may be held at a voltage greater than that at pin 14 or less than that at pin 7.

科技英语翻译知识

词类的转换

　　英汉两种语言在词汇的使用上有很大的不同。汉语里多使用动词，而英语（尤其是科技英语）里多使用名词；英语有许多名词派生的动词，以及由名词转用的动词，而汉语里没有；汉语里形容词可作谓语，而英语里形容词只能用作定语和表语等。因此在英译汉过程中，不可以逐词死译，而需要转换词性，才能使汉语译文通顺自然。

　　词性转换分为以下4种情况。

1. 转换成汉语动词

(1) Reversing the direction of the current reverses the direction of its lines of force.

倒转电流的方向也就倒转了它的磁力线的方向。

(动名词 reversing 转换成动词"倒转"。)

(2) It is desirable to transform high voltage into low voltage by means of a transformer in an AC system.

我们希望通过交流系统中的变压器把高压变为低压。

(形容词 desirable 转换成动词"希望"。)

(3) The electric current flows through the circuit with the switch on.

如果开关接通,电流就流过线路。

(副词 on 转换成动词"接通"。)

(4) The letter E is commonly used for electromotive force.

通常用字母 E 表示电动势。

(介词 for 转换成动词"表示"。)

2. 转换成汉语名词

(1) Neutrons act differently from protons.

中子的作用不同于质子。

(动词 act 转换成名词"作用"。)

(2) The electrons flow from the negative to the positive.

电子从负极流向正极。

(形容词 negative 转换成了名词"负极"。)

(3) Such magnetism, because it is electrically produced, is called electromagnetism.

由于这种磁性产生于电,因此称为电磁。

(副词 electrically 转换成名词"电"。)

(4) The voltage induced in the core is small, because it is essentially a winding having but one turn.

铁芯的感应电压小,因为铁芯实质上是一只有一匝的绕组。

(代词 it 转换成名词"铁芯"。)

3. 转换成汉语形容词

(1) Thus, the addition of the inductor prevents the current from building up or going down quickly.

这样,添加的电感器阻止电流不能迅速增加或下降。

(为与谓语"阻止"搭配,将定语"电感器"转译为主语,而将名词主语"添加"转换为"电感器"的定语,变为形容词。)

(2) Gases conduct best at low pressure.

气体在低压下导电性能最佳。

（副词 best 转换成形容词"最佳"。）

(3) The electric conductivity has great importance in selecting electrical materials.

导电性在选择电气材料时很重要。

（名词 importance 转换成形容词"重要"。）

(4) Why is electricity widely used in industry?

为什么电在工业中得到广泛的使用呢？

（动词 use 转换成名词，所以副词 widely 相应地转换成形容词"广泛的"。）

4. 转换成汉语副词

(1) The potentials of both the grid and plate are effective in controlling plate current.

栅极和屏极两者的电位有效地控制着屏极电流。

（形容词 effective 转译成副词"有效地"。）

(2) The added device will ensure accessibility for part loading and unloading.

增添这种装置将保证工件装卸方便。

（名词 accessibility 转换成副词"方便"修饰动词"装卸"。）

(3) In the long and laborious search, Marie Curie succeeded in isolating two new substances.

在长期辛勤的研究中，玛丽·居里成功地分解出两种新物质。

（动词 succeed 转译成副词"成功地"。）

(4) In human society activity in production develops step by step from a lower to higher level.

人类社会的生产活动，由低级向高级逐步发展。

（状语 step by step 转译成副词"逐步"修饰发展。）

Reading Material

阅读下列文章。

Text	Note
Embedded System An embedded system is a combination of computer hardware and software designed for a specific function. Embedded systems may also function within a larger system. The systems can be programmable or have a fixed functionality. Industrial machines, consumer electronics, agricultural[1] and processing industry devices, automobiles, medical equipment, cameras, digital watches, household appliances, airplanes, vending machines and toys as well as mobile devices are possible locations for an embedded system. While embedded systems are computing systems, they can range	[1] *adj.* 农业的

from having no user interface (UI) — for example, on devices designed to perform a single task — to complex graphical user interfaces (GUIs), such as in mobile devices. User interfaces can include buttons, LEDs (light-emitting diodes) and touchscreen[2] sensing. Some systems use remote user interfaces as well.

[2] *n.* 触摸屏

1. How does an Embedded System Work?

Embedded systems always function as part of a complete device—that's what's meant by the term embedded. They are low-cost, low-power-consuming, small computers that are embedded in other mechanical or electrical systems. Generally, they comprise[3] a processor, power supply, and memory and communication ports. Embedded systems use the communication ports to transmit data between the processor and peripheral devices—often, other embedded systems—using a communication protocol. The processor interprets this data with the help of minimal software stored on the memory. The software is usually highly specific to the function that the embedded system serves.

[3] *vt.* 包含，包括

The processor may be a microprocessor or microcontroller[4]. Microcontrollers are simply microprocessors with peripheral interfaces and integrated memory included. Microprocessors use separate integrated circuits for memory and peripherals instead of including them on the chip. Microprocessors typically require more support circuitry than microcontrollers because there is less integrated into the microprocessor. The term system on a chip (SoC) is often used. SoCs include multiple processors and interfaces on a single chip. They are often used for high-volume embedded systems. Some example are the application-specific integrated circuit (ASIC) and the field-programmable gate array (FPGA).

[4] *n.* 微控制器

Often, embedded systems are used in real-time operating environments and use a real-time operating system (RTOS) to communicate with the hardware. Near-real-time approaches are suitable at higher levels of chip capability, defined by designers who have increasingly decided the systems are generally fast enough and the tasks are tolerant[5] of slight variations in reaction. In these instances, stripped-down[6] versions of the Linux operating system are commonly deployed, although other OSes have been pared down to run on embedded systems, including Embedded Java and Windows IoT

[5] *adj.* 宽容的，容忍的
[6] *adj.* 精简的，简装的

(formerly Windows Embedded).

2. Characteristics of Embedded Systems

The main characteristic of embedded systems is that they are task-specific[7].

[7] *adj.* 任务特定的

Additionally, embedded systems can include the following characteristics:

(1) They typically consist of hardware, software and firmware;

(2) They can be embedded in a larger system to perform a specific function, as they are built for specialized tasks within the system, not various tasks;

(3) They can be either microprocessor-based or microcontroller-based—both are integrated circuits that give the system compute power;

(4) They are often used for sensing and real-time computing in internet of things (IoT) devices, which are devices that are internet-connected and do not require a user to operate;

(5) They can vary in complexity and in function, which affects the type of software, firmware[8] and hardware they use; and

[8] *n.* 固件

(6) They are often required to perform their function under a time constraint[9] to keep the larger system functioning properly.

[9] *n.* 限制，约束

3. Structure of Embedded Systems

Embedded systems vary in complexity but, generally, consist of three main elements:

(1) Hardware. The hardware of embedded systems is based around microprocessors and microcontrollers. Microprocessors are very similar to microcontrollers and, typically, refer to a CPU (central processing unit) that is integrated with other basic computing components such as memory chips and digital signal processors (DSPs). Microcontrollers have those components built into one chip.

(2) Software and firmware. Software for embedded systems can vary in complexity. However, industrial-grade[10] microcontrollers and embedded IoT systems usually run very simple software that requires little memory.

[10] *adj.* 工业级的

(3) Real-time operating system. These are not always included in embedded systems, especially smaller-scale systems. RTOSes define how the system works by supervising the software and setting rules

during program execution[11].

In terms of hardware, a basic embedded system would consist of the following elements:

(1) Sensors. They convert physical sense data into an electrical signal.

(2) Analog to digital converters. They change an analog electrical signal into a digital one.

(3) Processors. They process digital signals and store them in memory.

(4) Digital to analog converters. They change the digital data from the processor into analog data.

(5) Actuators. They compare actual output to memory-stored output and choose the correct one.

The sensor reads external inputs, the converters make that input readable to the processor, and the processor turns that information into useful output for the embedded system.

4. Embedded System Trends

While some embedded systems can be relatively simple, they are becoming more complex, and more and more of them are now able to either supplant human decision-making or offer capabilities beyond what a human could provide. For instance, some aviation[12] systems, including those used in drones, are able to integrate sensor data and act upon that information faster than a human could, permitting new kinds of operating features.

The embedded system is expected to continue growing rapidly, driven in large part by the internet of things. Expanding IoT applications, such as wearables, drones, smart homes, smart buildings, video surveillance[13], 3D printers[14] and smart transportation, are expected to fuel[15] embedded system growth.

[11] *n.* 实行,执行

[12] *n.* 航空

[13] *n.* 监控,监视

[14] *n.* 打印机;印刷机

[15] *v.* 刺激;加剧;加燃料 *n.* 燃料;刺激物

参 考 译 文

模拟电路和数字电路

模拟电路和数字电路都广泛应用于电气和电子工程的各个领域中的信号处理。

1. 什么是模拟电路？

模拟电路是一种可以处理任何模拟信号并以模拟形式产生输出的电子电路（见图 5-1）。模拟电路由电阻、电感、电容等组成。

（图略）

作为时间的连续函数的信号称为模拟信号。所有现实世界的信号都是模拟信号，因此，模拟电路不需要对输入信号进行任何转换，即模拟输入信号可以直接馈送到模拟电路，没有任何损失，并且可以直接由给定的模拟电路处理。此外，模拟电路产生的输出信号是模拟信号。

根据电路行为和所使用的元件，模拟电路可以分为两种类型：有源电路和无源电路。放大器是有源电路的示例，而低通滤波器是无源电路的示例。模拟电路的主要缺点是模拟信号非常容易受到噪声的影响，噪声可能会导致信号波形失真和信息丢失。

2. 什么是数字电路？

数字电路是处理数字信号的电子电路。作为时间的离散函数的信号称为数字信号。

数字电路的基本构建模块是数字逻辑门（见图 5-2）。数字电路只能处理数字信号，但现实世界的信号具有模拟性质。因此，需要使用被称为模数转换器（ADC）的特殊电子电路将它们转换为数字信号。数字电路的输出也是数字信号，需要将其转换回模拟信号。

（图略）

数字电路在采样过程中可能会丢失信息。数字电路只能是有源电路，这意味着它们需要额外的电源来为电路供电。

3. 模拟电路与数字电路的区别

表 5-1 重点介绍了模拟电路和数字电路之间的主要区别。

表 5-1　模拟电路和数字电路之间的主要区别

参　　数	模　拟　电　路	数　字　电　路
定义	只能处理模拟信号的电子电路称为模拟电路	只能处理数字信号的电子电路称为数字电路
输入信号	模拟电路的输入信号必须是连续时间信号或模拟信号	数字电路的输入信号是离散时间信号或数字信号
输出信号	模拟电路以模拟信号的形式产生输出	数字电路以数字信号的形式产生输出
电路元件	模拟电路的电路元件有电阻、电感、电容等	数字电路的主要电路元件是逻辑门
是否需要转换器	模拟电路可以直接处理自然界中存在的模拟信号。因此，模拟电路不需要信号转换器	数字电路只能处理数字形式的信号。因此，数字电路需要信号转换器，即模数转换器（ADC）和数模转换器（DAC）
对噪声的敏感性	模拟信号更容易受到噪声的影响	数字信号不受噪声影响
设计	模拟电路设计复杂，因为其电路元件需要手动放置	通过使用多种软件，复杂的数字电路的设计相对容易
灵活性	模拟电路的实现不灵活	数字电路的实现过程更灵活

续表

参　　数	模 拟 电 路	数 字 电 路
类型	模拟电路可以有两种类型：有源电路和无源电路	数字电路只有一种类型，称为有源电路
处理速度	模拟电路的处理速度相对较低	数字电路比模拟电路具有更高的处理速度
电耗	模拟电路消耗更多电量	数字电路消耗的电量相对较少
准确度和精确度	模拟电路的准确度和精确度较差	数字电路相对更准确和精确
观测误差	模拟电路的输出中可能存在观测误差	数字电路的输出没有观测误差
信号传输	模拟电路中，信号以波的形式无线或有线传输	数字电路中，信号只能以数字形式通过导线传输
信息存储形式	模拟电路以波的形式存储信息	数字电路以二进制形式存储信息
逻辑运算	模拟电路无法有效地执行逻辑运算	数字电路能够有效地执行逻辑运算

4. 模数转换器（ADC）

ADC 是现代电子设备的重要组件。它们用于将模拟信号（如声音或光）转换为可以由计算机和其他数字设备处理的数字数据。ADC 应用广泛，从音频录制和播放到医疗成像和工业控制系统。

ADC 的工作原理是将模拟信号转换为一系列数字值（见图 5-3）。将模拟信号转换为数字数据的过程分为三个步骤，包括采样、量化和编码。

（图略）

4.1　采样

转换过程的第一步是采样。它包括以固定速率对模拟信号进行定期采样。采样速率称为采样率，以每秒采样数来衡量。采样率越高，采到的样本就越多，模拟信号的数字表示就越准确。

采样是转换过程中的关键步骤，因为它决定了模拟信号的数字表示的准确性。如果采样率太低，模拟信号的数字表示将不准确，并且可能会丢失信号的重要细节。

4.2　量化

转换过程的第二步是量化。量化指为模拟信号的每个样本分配一个数字值。量化过程是一个离散步骤，它将输入电压分为数量有限的若干步骤。每个步骤都分配一个数字值，它代表该时间点输入信号的电压。量化过程中的步骤越多，模拟信号的数字表示就越准确。

量化确保了模拟信号的数字表示是准确和精确的。如果没有量化，模拟信号的数字表示将是连续的，并且不可能使用数字设备存储或处理信号。

4.3　编码

转换过程的最后一步是编码。编码指将量化过程产生的数字值转换为可以由计算机和其他数字设备存储和处理的二进制代码。最常见的编码方法是二进制代码，其中每个数字值都由一串 1 和 0 表示。

编码把模拟信号表示为数字,以便计算机和其他数字设备进行存储和处理。如果没有编码,模拟信号的数字表示对于数字设备来说毫无意义,并且不可能将该信号用于任何实际目的。

5. 数模转换器(DAC)

DAC 是将数字信号转换为模拟信号的设备(见图 5-4)。它常用于各种电子系统,如音频播放器、数字乐器和通信设备。将数字信号转换为模拟信号的过程涉及几个步骤。
(图略)

5.1 输入数字信号

DAC 接收数字信号作为输入。该数字信号通常采用二进制数据的形式,其中每一位代表一个离散值。

5.2 采样保持

数字信号首先通过采样保持电路。该电路定期对数字信号进行采样,并保存每个采样值,直到进行下一个采样。此过程有助于从离散数字样本重建连续时间模拟信号。

5.3 转换

然后将采样的数字值馈送到数模转换电路。该转换电路的目的是将离散的数字值转换为连续的模拟电压或电流。DAC 架构有多种类型,如电阻梯形 DAC、delta-sigma DAC 和乘法 DAC,每种架构都有自己的实现细节。然而,基本原理是使用数字输入来生成相应的模拟输出。

5.4 重建滤波器

DAC 的输出是模拟信号,由于采样过程,因此可能会包含一些不需要的高频分量。为了去除这些不需要的成分,通常使用重建滤波器。重建滤波器衰减高频分量并重建平滑的模拟信号。

5.5 输出放大

在某些情况下,DAC 的输出信号可能需要放大才能达到所需的信号电平。放大器用于在将模拟信号发送到输出设备(如扬声器或显示器)之前增加模拟信号的功率或电压电平。

Text A

Electronic Circuit Analysis and Design

Electronic circuit analysis and design is a crucial aspect of electrical engineering. It involves the study of electronic circuits and their behavior under different conditions. The analysis and design of electronic circuits are essential for the development of new technologies and the improvement of existing ones.

1. Overview of Electronic Circuit Analysis and Design

1.1 Circuit analysis

Circuit analysis involves understanding how electronic circuits work. This includes understanding the behavior of components such as resistors, capacitors, and inductors, as well as understanding how to analyze circuits using techniques such as Kirchhoff's laws and nodal analysis.

1.2 Circuit design

Circuit design involves creating electronic circuits to meet specific requirements. This includes selecting components, designing circuits to meet performance specifications, and testing and refining the circuit design.

2. Circuit Analysis Techniques

2.1 Kirchhoff's laws

Kirchhoff's laws are fundamental laws that govern the behavior of electrical circuits. The first law, also known as Kirchhoff's current law (KCL), states that the sum of currents entering a node is equal to the sum of currents leaving the node.[1] The second law, also known as Kirchhoff's voltage law (KVL), states that the sum of voltages around any closed loop in a circuit is equal to zero. These laws are essential for analyzing complex circuits and can be used to determine the values of currents and voltages at different points in a circuit.

2.2 Nodal analysis

Nodal analysis is a technique used to determine the voltages at different nodes in a circuit. It involves applying KCL at each node and solving a system of equations to obtain the voltage values. This technique is particularly useful for circuits with many nodes and can be used to find the voltage at any node in the circuit.

2.3 Mesh analysis

Mesh analysis is a technique used to determine the currents flowing through different loops in a circuit. It involves applying KVL around each loop and solving a system of equations to obtain the current values. This technique is particularly useful for circuits with many loops and can be used to find the current flowing through any loop in the circuit.

2.4 Thevenin's and Norton's theorems

Thevenin's and Norton's theorems are techniques used to simplify complex circuits into simpler equivalent circuits. Thevenin's theorem states that any linear circuit can be replaced by an equivalent circuit consisting of a voltage source in series with a resistor.[2] Norton's Theorem states that any linear circuit can be replaced by an equivalent circuit consisting of a current source in parallel with a resistor. These theorems are particularly useful for simplifying complex circuits and can be used to determine the behavior of a circuit under different conditions.

2.5 Transient analysis

Transient analysis is a technique used to analyze the behavior of a circuit during the transient period, which is the period immediately following a change in the circuit.[3] This technique involves solving differential equations to obtain the voltage and current values at different points in the circuit. Transient analysis is particularly useful for circuits with capacitors and inductors as these components can cause significant changes in the circuit during the transient period.

In conclusion, these circuit analysis techniques are essential for analyzing and designing complex electrical circuits. By applying these techniques, engineers can determine the behavior of a circuit under different conditions and design circuits that meet specific requirements.

3. Circuit Components

3.1 Passive circuit components

Passive circuit components are those that do not require an external source of power to operate. These include resistors, capacitors, and inductors. These components are used to create circuits that can perform a variety of tasks, such as filtering, amplification, and signal processing.[4]

Resistors are used to limit the flow of current in a circuit. They are commonly used in voltage dividers and current-limiting circuits. Capacitors are used to store charge and are commonly used in filter circuits to remove unwanted frequencies from a signal. Inductors are used to store energy in a magnetic field and are commonly used in filter circuits and power supplies.

3.2　Active circuit components

Active circuit components are those that require an external source of power to operate. These include transistors, diodes, and operational amplifiers. These components are used to create circuits that can perform a variety of tasks, such as amplification, switching, and signal processing.

Transistors are used to amplify and switch signals. They are commonly used in amplifiers, oscillators, and switching circuits. Diodes are used to rectify AC signals and are commonly used in power supplies. Operational amplifiers are used to amplify and process signals and are commonly used in filter circuits and amplifiers.

3.3　Amplifiers

Amplifiers are used to increase the amplitude of a signal. There are many different types of amplifiers. The choice of amplifier type depends on the application and the desired performance characteristics.

3.4　Filters

Filters are used to remove unwanted frequencies from a signal. There are many different types of filters, including low-pass, high-pass, band-pass, and band-stop filters. The choice of filter type depends on the application and the desired frequency response.

3.5　Oscillators

Oscillators are used to generate a periodic waveform. There are many different types of oscillators, including RC oscillators, LC oscillators, and crystal oscillators. The choice of oscillator type depends on the application and the desired frequency stability.

In conclusion, understanding the circuit components is essential for creating circuits that can perform a variety of tasks. By carefully selecting passive and active components, choosing the appropriate amplifiers, filters, and oscillators, it is possible to create circuits that meet the requirements of a wide range of applications.[5]

4. Expertise Required in Advanced Electronic Circuit Analysis and Design

4.1　Nonlinear circuit analysis

Nonlinear circuit analysis is an essential aspect of electronic circuit design. It involves the study of circuits that exhibit nonlinear behavior, such as diodes, transistors, and operational amplifiers. Nonlinear circuits are crucial in modern electronic devices, and their analysis and design require advanced mathematical techniques and simulation tools.

4.2 Digital circuit analysis

Digital circuit analysis is the study of circuits that process digital signals. It involves the design and analysis of logic gates, flip-flops, counters, and other digital components. Digital circuits are used in a wide range of electronic devices, such as computers, smartphones, and digital cameras.

4.3 Signal processing circuits

Signal processing circuits are used to modify, filter, and amplify analog signals. They are essential in communication systems, audio and video processing, and instrumentation. Signal processing circuits require advanced analysis and design techniques, such as Fourier analysis, Laplace transforms, and filter design.

4.4 RF circuit design

RF circuit design is the study of circuits that operate at radio frequencies. It involves the design and analysis of amplifiers, mixers, filters, and antennas. RF circuits are used in communication systems, radar, and wireless devices.

4.5 Power electronics

Power electronics is the study of circuits that convert and control electrical power. It involves the design and analysis of power supplies, motor drives, and power converters. Power electronics is essential in renewable energy systems, electric vehicles, and industrial automation.

In conclusion, advanced electronic circuit analysis and design require expertise in various areas, such as nonlinear circuit analysis, digital circuit analysis, signal processing circuits, RF circuit design, and power electronics. By understanding these areas, engineers can design and analyze complex electronic circuits that meet the requirements of modern electronic devices.

New Words

nodal	['nəudəl]	adj.	节点的
refine	[rɪ'faɪn]	vt.	改善；提炼
state	[steɪt]	vt.	规定；陈述，声明
obtain	[əb'teɪn]	vt.	获得，得到
loop	[luːp]	n.	环，圈
linear	['lɪnɪə]	adj.	线性的；[数]一次的
significant	[sɪg'nɪfɪkənt]	adj.	显著的；重要的
engineer	[ˌendʒɪ'nɪə]	n.	工程师
oscillator	['ɒsɪleɪtə]	n.	振荡器
amplitude	['æmplɪtjuːd]	n.	振幅
low-pass	[ləʊ pɑːs]	adj.	低通的
high-pass	[haɪ pɑːs]	adj.	高通的

band-pass	[bænd pɑːs]	*adj.* 带通的
band-stop	[bænd stɒp]	*adj.* 带阻的
periodic	[ˌpɪəriˈɒdɪk]	*adj.* 周期的；定期的
waveform	[ˈweɪvfɔːm]	*n.* 波形
exhibit	[ɪgˈzɪbɪt]	*vt.* 呈现
flip-flop	[flɪp flɒp]	*n.* 触发器
smartphone	[ˈsmɑːtfəʊn]	*n.* 智能手机
modify	[ˈmɒdɪfaɪ]	*vi.* 修改
instrumentation	[ˌɪnstrəmenˈteɪʃn]	*n.* 仪器仪表

Phrases

electronic circuit analysis	电路分析
Kirchhoff's law	基尔霍夫定律
nodal analysis	节点分析
circuit design	电路设计
closed loop	闭环
complex circuit	复杂电路
system of equation	方程组
mesh analysis	网格分析
Thevenin's theorem	戴维南定理
Norton's theorem	诺顿定理
simplify…into…	把……简化为……
equivalent circuit	等效电路
linear circuit	线性电路
be replaced by…	被……替换
voltage source	电压源
in series with…	与……串联，与……相连
current source	电流源
transient analysis	暂态分析，瞬态分析
passive circuit component	无源电路元件
external source of power	外部电源
voltage divider	分压器
current-limiting circuit	限流电路
filter circuit	滤波电路
power supply	电源，供电
active circuit component	有源电路元件
AC signal	交流信号
operational amplifier	运算放大器
frequency response	频率特性；频率响应

crystal oscillator	晶体振荡器
nonlinear circuit analysis	非线性电路分析
digital circuit analysis	数字电路分析
logic gate	逻辑门，逻辑闸
digital camera	数码相机
Fourier analysis	傅里叶分析
Laplace transform	拉普拉斯变换
radio frequency	射频
wireless device	无线电设备
power electronic	电力电子
power converter	整流器
renewable energy system	可再生能源系统

Abbreviations

KCL (Kirchhoff's Current Law)	基尔霍夫电流定律
KVL (Kirchhoff's Voltage Law)	基尔霍夫电压定律
RC (Resistor-Capacitance)	电阻-电容
LC (inductance-Capacitance)	电感-电容
RF (Radio Frequency)	射频

Notes

[1] The first law, also known as Kirchhoff's current law (KCL), states that the sum of currents entering a node is equal to the sum of currents leaving the node.

本句中，also known as Kirchhoff's current law (KCL)对主语 The first law 进行补充说明。that the sum of currents entering a node is equal to the sum of currents leaving the node 是一个宾语从句，作 states 的宾语。在该宾语从句中，entering a node 是一个现在分词短语，作定语，修饰和限定从句的主语，即它前面的 the sum of currents。leaving the node 也是一个现在分词短语，作定语，修饰和限定它前面的 the sum of currents。is equal to 的意思是"等于"。

本句意为：第一定律，也称为基尔霍夫电流定律(KCL)，表明进入节点的电流总和等于离开节点的电流总和。

[2] Thevenin's theorem states that any linear circuit can be replaced by an equivalent circuit consisting of a voltage source in series with a resistor.

本句中，Thevenin's theorem 是主语，states 是谓语，that any linear circuit can be replaced by an equivalent circuit consisting of a voltage source in series with a resistor 是一个宾语从句，作 states 的宾语。在该宾语从句中，consisting of a voltage source in series with a resistor 是一个现在分词短语，作定语，修饰和限定 an equivalent circuit。be replaced by 的意思是"被……代替"。

本句意为：戴维南定理指出，任何线性电路都可以用由电压源与电阻器串联组成的等效电路代替。

[3] Transient analysis is a technique used to analyze the behavior of a circuit during the transient period, which is the period immediately following a change in the circuit.

本句中，used to analyze the behavior of a circuit during the transient period 是一个过去分词短语，作定语，修饰和限定 a technique。which is the period immediately following a change in the circuit.是一个非限定性定语从句，对 the transient period 进行补充说明。

本句意为：瞬态分析是一种用于分析瞬态期间电路行为的技术，瞬态期间是电路发生变化后的一段时间。

[4] These components are used to create circuits that can perform a variety of tasks, such as filtering, amplification, and signal processing.

本句中，to create circuits that can perform a variety of tasks 是一个动词不定式短语，作目的状语，修饰谓语 are used。在该短语中，that can perform a variety of tasks 是一个定语从句，修饰和限定 circuits。such as filtering, amplification, and signal processing 是对 a variety of tasks 的举例说明。

本句意为：这些元件用于创建可以执行各种任务的电路，如滤波、放大和信号处理。

[5] By carefully selecting passive and active components, choosing the appropriate amplifiers, filters, and oscillators, it is possible to create circuits that meet the requirements of a wide range of applications.

本句中，By carefully selecting passive and active components, choosing the appropriate amplifiers, filters, and oscillators 作方式状语。it 是形式主语，真正的主语是动词不定式短语 to create circuits that meet the requirements of a wide range of applications。that meet the requirements of a wide range of applications 是一个定语从句，修饰和限定 circuits。meet the requirements of 的意思是"满足……的要求"，a wide range of 的意思是"各种各样的，广泛的，很多"。

本句意为：通过仔细选择无源元件和有源元件，以及适当的放大器、滤波器和振荡器，可以产生满足各种应用要求的电路。

Exercises

【Ex.1】根据课文内容，回答以下问题。

1. What does circuit analysis involve? What does it include?

2. What does circuit design involve? What does it include?

3. What is nodal analysis? What does it involve?

4. What are passive circuit components and active circuit components respectively?

5. What are the expertise required in advanced electronic circuit analysis and design?

【Ex.2】根据下面的英文解释，写出相应的英文词汇。

英 文 解 释	词　　汇
a complete electrical circuit around which current flows or a signal circulates	
a piece of equipment for producing oscillating electric currents	
a curve showing the shape of a wave at a given time	
a collective term for measuring instruments, used for indicating, measuring and recording physical quantities	
the term which refers to creating electronic circuits, it can range from individual transistors in an integrated circuit to complex circuits	
a technique for calculating currents in planar circuits at any point along the circuit	
an electronic component which can only receive energy, which it can either dissipate, absorb or store it in an electric field or a magnetic field	
the element or device which is capable of providing or delivering energy to the circuit	
a device performing a Boolean logic operation on one or more binary inputs and then outputs a single binary output	
any type of communications device that does not require a physical wire to relay information	

【Ex.3】把下列句子翻译成中文。

1. The amplitude of the vibration determines the loudness of the sound.

2. Kirchhoff's current law (1st Law) states that the current flowing into a node (or a junction) must be equal to the current flowing out of it.

3. Thevenin's theorem is a fundamental principle in electrical circuit analysis.

4. Norton's theorem states that any linear circuit can be simplified to an equivalent circuit consisting of a single current source and parallel resistance that is connected to a load.

5. A linear circuit always consists of linear elements, linear independent and dependent sources.

6. A nonlinear circuit consists of at least one nonlinear element, not counting the voltage and current independent sources.

7. Transient analysis is all about determining how a circuit responds to changes in the driving voltage/current.

8. A filter circuit typically filters out the peaks and valleys of an electrical signal.

9. FFT (Fast Fourier Transform) is one of the most useful analysis tools available.

10. A digital circuit is a type of circuit that operates on different logic gates.

【Ex.4】把下列短文翻译成中文。

Linear Circuit and Non-linear Circuit

1. What Is a Linear Circuit?

An electrical circuit in which the value of parameters or elements remains constant is said to be known as a linear circuit. In other words, the parameters of a linear circuit don't change with respect to the voltage and current in the circuit.

2. What Is a Non-linear Circuit?

An electrical circuit in which the value of parameters or elements does not remain constant and varies with the applied voltage and current is called a non-linear circuit. In non-linear circuits, the curve between the current through it and the voltage across it do not result in a straight line (non-linear).

3. Key Differences Between Linear and Non-linear Circuits:

(1) In a linear circuit, the parameters (resistance, inductance, capacitance, etc.) are always constant irrespective of variations in current or voltage. Whereas in a non-linear circuit, the parameters change with voltage or current.

(2) The current is directly proportional to voltage in a linear circuit, but in a non-linear circuit, the characteristics between current and voltage are non-linear.

(3) A linear circuit consists of only linear elements, whereas a non-linear circuit consists of

at least one nonlinear element.

(4) A linear circuit obeys the properties of superposition (additive) and homogeneity (scaling). Whereas a non-linear circuit doesn't obey any of these properties.

【Ex.5】通过 Internet 查找资料，借助电子词典、辅助翻译软件及 AI 工具，完成以下技术报告，并附上收集资料的网址。通过 E-mail 发送给老师，或按照教学要求在网上课堂提交。
1. 当前有哪些最主要的电路分析软件并简述各个软件的主要功能。
2. 当前有哪些最主要的电路设计软件并简述各个软件的主要功能。

Text B

Basic DC Motor Operation

1. Magnetic Fields

As is known that there are two electrical elements of a DC motor, the field windings and the armature. The armature windings are made up of current carrying conductors that terminate at a commutator. DC voltage is applied to the armature windings through carbon brushes which ride on the commutator. In small DC motors, permanent magnets can be used for the stator. However, in large motors used in industrial applications the stator is an electromagnet. When voltage is applied to stator windings an electromagnet with north and south poles is established. The resultant magnetic field is static (nonrotational). For simplicity of explanation, the stator will be represented by permanent magnets in the following illustrations (see Figure 6-1).

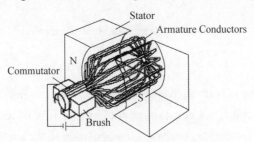

Figure 6-1　Structure of DC motor

A DC motor rotates as a result of two magnetic fields interacting with each other (see Figure 6-2). The first field is the main field that exists in the stator windings. The second field exists in the armature. Whenever current flows through a conductor a magnetic field is generated around the conductor.

2. Right-Hand Rule for Motors

A relationship, known as the right-hand rule for motors, exists between the main field, the field around a conductor, and the direction the conductor tends to move. If the thumb, index finger,

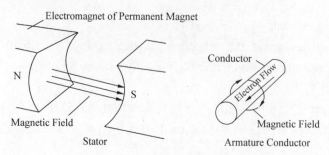

Figure 6-2　Magnetic fields of DC motors

and third finger are held at right angles to each other and placed as shown in the following illustration so that the index finger points in the direction of the main field flux and the third finger points in the direction of electron flow in the conductor, the thumb will indicate direction of conductor motion. As can be seen from the following illustration, conductors on the left side tend to be pushed up (see Figure 6-3). Conductors on the right side tend to be pushed down. This results in a motor that is rotating in a clockwise direction. You will see later that the amount of force acting on the conductor to produce rotation is directly proportional to the field strength and the amount of current flowing in the conductor.

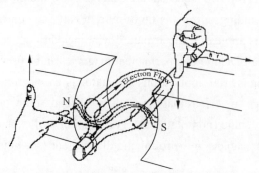

Figure 6-3　The right-hand rule for motors

3. CEMF

Whenever a conductor cuts through lines of flux a voltage is induced in the conductor. In a DC motor the armature conductors cut through the lines of flux of the main field. The voltage induced into the armature conductors is always in opposition to the applied DC voltage. Since the voltage induced into the conductor is in opposition to the applied voltage it is known as CEMF (Counter Electromotive Force) (see Figure 6-4). CEMF reduces the applied armature voltage.

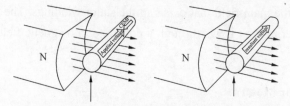

Figure 6-4　Counter Electromotive Force

The amount of induced CEMF depends on many factors such as the number of turns in the coils, flux density, and the speed which the flux lines are cut.

4. Armature Field

An armature, as we have learned, is made up of many coils and conductors. The magnetic fields of these conductors combine to form a resultant armature field with a north and a south poles. The north pole of the armature is attracted to the south pole of the main field. The south pole of the armature is attracted to the north pole of the main field. This attraction exerts a continuous torque on the armature. Even though the armature is continuously moving, the resultant field appears to be fixed (see Figure 6-5). This is due to commutation.

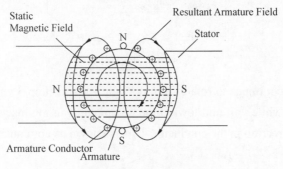

Figure 6-5　Armature field

5. Commutation

In the following illustration of a DC motor only one armature conductor is shown (see Figure 6-6). Half of the conductor has been shaded black, the other half white. The conductor is connected to two segments of the commutator. In position 1 the black half of the conductor is in contact with the negative side of the DC applied voltage. Current flows away from the commutator on the black half of the conductor and returns to the positive side, flowing towards the commutator on the white half.

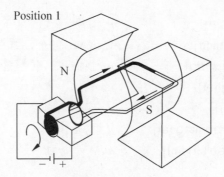

Figure 6-6　Position 1

In position 2 the conductor has rotated 90° (see Figure 6-7). At this position the conductor is lined up with the main field. This conductor is no longer cutting main field magnetic lines of flux;

therefore, no voltage is being induced into the conductor. Only applied voltage is present. The conductor coil is short-circuited by the brush spanning the two adjacent commutator segments. This allows current to reverse as the black commutator segment makes contact with the positive side of the applied DC voltage and the white commutator segment makes contact with the negative side of the applied DC voltage.

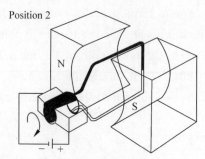

Figure 6-7   Position 2

As the conductor continues to rotate from position 2 to position 3 current flows away from the commutator in the white half and toward the commutator in the black half (see Figure 6-8). Current has reversed direction in the conductor. This is known as commutation.

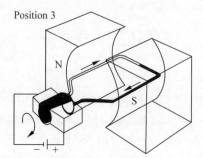

Figure 6-8   Position 3

New Words

winding	['waɪndɪŋ]	n. 绕组，线圈
terminate	['tɜːmɪneɪt]	vt. & vi. 停止，结束，终止
commutator	['kɒmjuteɪtə]	n. 换向器，转接器
magnet	['mæɡnət]	n. 磁体，磁铁
stator	['steɪtə]	n. 定子，固定片
electromagnet	[ɪ'lektrəʊmæɡnət]	n. 电磁体
nonrotational	['nɒnrəʊ'teɪʃənl]	adj. 无旋的
thumb	[θʌm]	n. 拇指
coil	[kɔɪl]	n. 线圈，绕组
flux	[flʌks]	n. 流量，通量
		vi. 熔化，流出

		vt. 使熔融
opposition	[ˌɒpəˈzɪʃn]	n. 相反；反对，对抗
resultant	[rɪˈzʌltənt]	adj. 结果必然产生的；[物]组合的，合成的
commutation	[ˌkɒmjuˈteɪʃn]	n. 转换，交换

Phrases

DC motor	直流电动机
field winding	磁场绕组
carbon brush	碳刷
resultant magnetic field	合成磁场
permanent magnet	永久磁铁
right-hand rule	右手定则
index finger	食指
third finger	中指
armature field	电枢场
push up	推上去，提高；顶
south pole	南极
north pole	北极
field strength	磁场强度
magnetic lines	磁力线
flux density	通量密度
be induce into	导入
in opposition to	与……相反,与……相对

Abbreviations

CEMF (Counter Electromotive Force)   反电动势

Exercises

【Ex.6】根据文章所提供的信息判断正误。

1. There are two electrical elements of a DC motor, the field windings and the stator.
2. In all DC motors, permanent magnets can be used for the stator.
3. Because two magnetic fields interact with each other, a DC motor works.
4. If you know the direction of electron flow in the conductor and the direction of the main field flux, you can decide the direction of conductor motion according to the right-hand rule.
5. If a conductor cuts through lines of flux, the voltage in the conductor is caused.
6. Whenever a conductor cuts through lines of flux a voltage is induced in the conductor. The voltage is CEMF.
7. The north pole of the armature is attracted to the south pole of the main field. The south pole of the armature is attracted to the north pole of the main field.

8. When the armature is moving, the resultant armature field will be changed.
9. The commutator can keep armature field to be fixed when armature is moving.
10. The commutator can change current direction in the conductor.

科技英语翻译知识

否定的译法

在英汉翻译，特别是科技英语翻译中，否定含义的理解和表达不容忽视。英语里表示否定意义的词汇、语法手段与汉语有很大不同。英语里的否定句有全部否定与部分否定之分；也有字面结构为肯定句，但实际含有否定意义的；还有实为强烈肯定的双重否定。翻译中必须正确理解才能正确表达。

1. 全部否定

Not、no、never、none、neither…nor、nothing、nowhere 等否定意义强烈的词出现在句中时，句子表示完全否定。翻译这类句子时，把表示否定的"不、无、非"等词与动词连用，即可达到全部否定。

(1) Rubber does not conduct electricity.

橡胶不导电。

(2) None of these metals have conductivity higher than copper.

这些金属中没有一种其电导率比铜高。

(3) A proton has a positive charge and an electron a negative charge, but a neutron has neither.

质子带正电荷，电子带负电荷，但中子不带任何电荷。

(4) A semiconductor is a material that is neither a good conductor nor a good insulator.

半导体材料既不是良导体，也不是好的绝缘体。

(5) There is no steel not containing carbon.

没有不含碳的钢。

2. 部分否定

由 all、every、both、always 等含有全体意义的词与否定词 not 构成部分否定。翻译时要特别注意，因为这样的搭配看似全部否定，而实际却是部分否定。

(1) All these metals are not good conductors.

这些金属并不都是良导体。

(2) Not all substances are good conductors of electricity.
并非所有的物质都是电的良导体。

(3) Both of the instruments are not precise.
这两台仪器不都是精密的。

(4) Every machine here is not imported from abroad.
这里的机器并非每台都是从国外进口的。

(5) But friction is not always useless, in certain cases it becomes a helpful necessity.
可是摩擦并不总是没用的，在某些情况下是有益而且必需的。

3. 意义否定
有些英语句子，在结构上是肯定句，因为包含有带否定意义的词或词组，实际却是否定句。

(1) The slight amount of current of flowing through the voltmeter may be ignored.
通过电压表的微量电流是可以忽略不计的。

(2) If the lighting or supply to other services fails, the battery supply takes over.
假如照明或别的供电失灵，蓄电池将承担供电任务。

(3) The purpose of the fuse is to protect such equipment from excessive current.
保险丝的用途是保护上述设备不受过量电流的影响。

(4) Another advantage of the absence of moving parts is that a transformer needs very little attention.
变压器没有运转部件的另一好处在于它几乎不需要关注。

(5) The atom is left deficient by one unit of negative charge and becomes a positive ion.
该原子失去一个单位的负电荷而变为正离子。

4. 双重否定
两个否定意义的词连用构成双重否定。这种否定实际是语气强烈的肯定，翻译时既可译成双重否定，也可转译成肯定句。

(1) No machines in the modern factories can work without electricity.
在现代工厂中，机器没有电就不能工作。

(2) Indeed, it is possible that electrons are made of nothing else but negative electricity.
的确，电子只可能是由负电荷构成的。

(3) A radar screen is not unlike television screen.
雷达荧光屏跟电视荧光屏没什么不一样。

(4) There has not been a scientist of eminence but was a man of industry.
没有哪一个有成就的科学家不是勤奋的。

(5) There is no rule that has no exceptions.
没有无例外的规则。

Reading Material

阅读下列文章。

Text	Note
Power Grid The power grid delivers electricity from power plants to homes and businesses across the nation. Its vast[1] network of power generation, transmission, and delivery ensures we can function in the modern world. The electrical grid provides us with electrical power on demand. **1. What Is a Power Grid and Its Function?** Whether you call it the power grid, power distribution grid, electrical grid, or national grid, this electrical network generates and distributes electricity across a large area. The network includes energy utility companies and energy suppliers that deliver electricity to your home or business. The power grid also consists of the infrastructure to generate and distribute power. The power grid does three things: It ensures best practice use of energy resources, provides greater power supply capacity, and makes power system operations more economical[2] and reliable. The generating stations are interconnected to reduce the reserve[3] generation capacity, known as a spinning reserve, in each area. **2. How Does an Electric Grid Work?** An electrical grid is a complex power generation, transmission, and distribution network. Grid operators — the entities that manage energy production and delivery — are regional[4] entities that control electrical energy as it travels through a fixed infrastructure. That infrastructure consists of power stations, transmission lines, and distribution lines. Sometimes it is called system operators or balancing authorities and grid operators manage the power grid to deliver your electricity. Grid operators monitor the power grid, signal to power plants when more power is needed and maintain the power grid's electrical flow to the transmission lines and distribution network.	[1] *adj.* 巨大的，大量的 [2] *adj.* 经济的，合算的 [3] *n. & vt.* 储备，保留 [4] *adj.* 地区的，区域的

A power grid has three functions: generation, transmission, and distribution. Within each step, complex processes are at work.

3. How Does a Power Plant Produce Electricity?

Your electric utility company's power plant generates electricity from three types of energy resources:

(1) Fossil[5] fuels, such as natural gas and coal;

(2) Nuclear[6] power;

(3) Renewable energy sources, including solar, wind, and hydropower[7].

Power plant output is measured in megawatts, and a megawatt (MW) is one million watts. Electrical output, the amount of electricity generated, varies depending on the size of a power plant. The average coal-fired[8] plant generates about 750MW of electricity.

Power plants use energy sources to generate electricity. Turbine generators produce electricity at the power plant using fuels from one of the three classes above. Wind turbines and solar photovoltaic[9] generation system can also produce electricity. Respectively, they use kinetic[10] energy and chemical energy to generate electricity.

To make the turbine blades[11] spin and rotate, a turbine generator works by pushing fluid, whether water, steam, air, or combustion[12] gases. Next, the turbine's rotor shaft[13], connected to a generator, converts the rotor's kinetic energy (also known as mechanical energy) to electrical energy.

4. What Do Transformers Do?

A transformer is an electrical device that moves electrical energy from one electric circuit to another using the principles of electromagnetic induction. Transformers don't generate electricity. They only transfer power from one alternating current to another.

Transformers can increase or decrease AC voltage, which is called stepping up or stepping down the voltage. For example, when the electricity leaves the power plant, it passes through a transformer to step up the voltage.

5. Why Are High-Voltage Transmission Lines Necessary?

High-voltage transmission lines carry high-voltage electricity over long distances, and they are instrumental[14] in delivering

[5] *n.* 化石 *adj.* 化石的
[6] *adj.* 原子核的,原子能的
[7] *n.* 水电

[8] *adj.* 烧煤的

[9] *adj.* 光伏的

[10] *adj.* 动力（学）的，运动的
[11] *n.* 叶片
[12] *n.* 燃烧
[13] *n.* 轴

[14] *adj.* 起作用的；有帮助的

electricity to the power grid's distribution networks. These high-voltage power lines carry up to 500,000 volts. A large industrial plant might also require high-voltage lines directly from overhead transmission lines. Without high-voltage transmission lines, the complexity and vulnerability of the power grid become more expensive and difficult to manage.

Power is conveyed[15] at a high voltage to increase efficiency by preventing energy loss. Higher voltage means lower current. In turn, lower current decreases resistance loss in conductors. Decreased resistance means less lost energy as electricity moves over long distances, making transmission lines an essential part of the power grid.

There are two kinds of electrical lines. High-voltage lines carry power from the power plants to substations[16], often over long distances. From the substation, distribution lines send energy to residential and business consumers. Distribution lines have a much lower voltage.

6. What Are Substations and What Do They Do?

When electricity from high-voltage transmission lines comes into a substation, transformers step down the voltage for distribution to homes and businesses. This happens at substations.

Substations are equipped with fuses[17] that split the current into multiple distribution lines. Through circuit breakers, switches, and capacitors, grid operators can also isolate and control the interface between high-voltage power lines and distribution lines.

When the electricity supply leaves the substation via the distribution lines, it's ready for delivery to homes and businesses.

7. What Causes Power Grid Failure?

A single event doesn't cause a power grid blackout[18]. Grid failure can result from a series of accidents[19] and missteps[20] that culminate in cascading failure. Cascading failure occurs when an element of the system partially or wholly fails. The collapse[21] creates a power surge load on other nearby system elements, which are stressed beyond capacity, causing them to fail.

For example, a power grid can fail if the frequency plunges below the lower limit of the band[22] or shoots up beyond the upper limit. When that happens, transmission lines no longer accept the power supply, which means other grid constituents[23], including the power

[15] *vt.* 运送，输送

[16] *n.* 变电站，变电所

[17] *n.* 保险丝

[18] *n.* 停电
[19] *n.* 事故
[20] *n.* 失误
[21] *v. & n.* 倒塌，崩溃

[22] *n.* 频段，频带

[23] *n.* 组件

generation plants, go offline.

8. What Factors Threaten the Electric Grid?

Climate change-induced extreme weather events such as heatwaves, blizzards[24], or hurricanes[25] can impact the electric grid. For example, 96% of power outages (blackouts) are weather-related or natural disaster[26] events that damage the electricity distribution systems infrastructure.

Surprisingly, the growth in renewable power sources also threatens the grid. This is because solar and wind power can sometimes produce more electricity than that the grid operators incorporate into the power grid system.

The increasing age of the power grid's infrastructure is another threat to the grid. Metal fatigue[27] and equipment deterioration[28] increasingly play a role in the grid's aging process.

9. What Happens If an Electric Grid Goes Down?

When a portion of the electric grid goes down, somebody somewhere is without electricity. In most cases, it's a relatively simple matter of rerouting[29] electricity from another grid sector or even another grid interconnection[30].

Unfortunately, large-scale energy storage isn't possible because electricity is a binary commodity (you have it or you don't). However, many utilities use lithium-ion[31] batteries as a short-term energy storage solution.

[24] *n.* 暴风雪
[25] *n.* 飓风
[26] *n.* 灾难

[27] *n.* 疲劳
[28] *n.* 恶化，退化

[29] *vt.* 变更路径
[30] *n.* 互相连接

[31] *n.* 锂离子

参 考 译 文

电路分析与设计

电路分析与设计是电气工程的一个重要方面。它研究电路及其在不同条件下的行为。电路分析与设计对于新技术的开发和现有技术的改进至关重要。

1. 电子电路分析与设计概述

1.1 电路分析

电路分析涉及了解电路的工作原理。这包括了解电阻器、电容器和电感器等组件的行为，以及了解如何使用基尔霍夫定律和节点分析等技术来分析电路。

1.2 电路设计

电路设计涉及创建电子电路以满足特定要求。这包括选择组件、设计电路以满足性能规格及测试和改进电路设计。

2. 电路分析技术

2.1 基尔霍夫定律

基尔霍夫定律是控制电路行为的基本定律。第一定律，也称为基尔霍夫电流定律（KCL），表明进入节点的电流总和等于离开节点的电流总和。第二定律，也称为基尔霍夫电压定律（KVL），表明电路中任何闭环回路的电压总和为0。这些定律对于分析复杂电路至关重要，可用于确定电路中不同点的电流值和电压值。

2.2 节点分析

节点分析是一种用于确定电路中不同节点的电压的技术。它涉及在每个节点应用KCL并求解方程组以获得电压值。该技术对于具有许多节点的电路特别有用，并且可用于查找电路中任何节点的电压。

2.3 网格分析

网格分析是一种用于确定流经电路中不同环路的电流的技术。它涉及在每个环路中应用KVL并求解方程组以获得电流值。该技术对于具有许多环路的电路特别有用，可用于查找流过电路中任何环路的电流。

2.4 戴维南定理和诺顿定理

戴维南定理和诺顿定理是用于将复杂电路简化为更简单的等效电路的技术。戴维南定理指出，任何线性电路都可以用由电压源与电阻器串联组成的等效电路代替。诺顿定理指出，任何线性电路都可以用由电流源与电阻器并联组成的等效电路代替。这些定理对于简化复杂电路特别有用，并且可用于确定电路在不同条件下的行为。

2.5 瞬态分析

瞬态分析是一种用于分析瞬态期间电路行为的技术，瞬态期间是电路发生变化后的一段时间。该技术涉及求解微分方程以获得电路中不同点的电压值和电流值。瞬态分析对于具有电容器和电感器的电路特别有用，因为这些元件可能会在瞬态期间导致电路发生显著变化。

总之，这些电路分析技术对于分析和设计复杂电路至关重要。通过应用这些技术，工程师可以确定电路在不同条件下的行为，并设计满足特定要求的电路。

3. 电路元件

3.1 无源电路元件

无源电路元件是那些不需要外部电源即可工作的元件。这些元件包括电阻器、电容器和电感器。这些元件用于创建可以执行各种任务的电路，如滤波、放大和信号处理。

电阻器用于限制电路中的电流。它们通常用于分压器和限流电路。电容器用于存储电荷，通常用于滤波器电路以消除信号中不需要的频率。电感器用于在磁场中存储能量，通常用于滤波器电路和电源。

3.2 有源电路元件

有源电路元件是那些需要外部电源才能运行的元件。这些元件包括晶体管、二极管和运算放大器。这些元件用于创建可以执行各种任务的电路，如放大、开关和信号处理。

晶体管用于放大和开关信号。它们通常用于放大器、振荡器和开关电路。二极管用于整流交流信号，常用于电源。运算放大器用于放大和处理信号，常用于滤波电路和放大器。

3.3 放大器

放大器用于增加信号的振幅。有许多不同类型的放大器。放大器类型的选择取决于应用和所需的性能特征。

3.4 滤波器

滤波器用于去除信号中不需要的频率。有许多不同类型的滤波器，包括低通滤波器、高通滤波器、带通滤波器和带阻滤波器。滤波器类型的选择取决于应用和所需的频率响应。

3.5 振荡器

振荡器用于生成周期性波形。振荡器有许多不同类型，包括 RC 振荡器、LC 振荡器和晶体振荡器。振荡器类型的选择取决于应用和所需的频率稳定性。

总之，了解电路元件对于创建可执行各种任务的电路至关重要。通过仔细选择无源元件和有源元件，以及适当的放大器、滤波器和振荡器，可以创建满足各种应用要求的电路。

4. 高级电子电路分析和设计所需的专业知识

4.1 非线性电路分析

非线性电路分析是电子电路设计的一个重要方面。它研究非线性行为的电路，如二极管、晶体管和运算放大器。非线性电路在现代电子设备中至关重要，其分析和设计需要先进的数学技术和仿真工具。

4.2 数字电路分析

数字电路分析研究处理数字信号的电路。它涉及逻辑门、触发器、计数器和其他数字元件的设计和分析。数字电路广泛应用于计算机、智能手机和数码相机等电子设备。

4.3 信号处理电路

信号处理电路用于修改、过滤和放大模拟信号。它们在通信系统、音频和视频处理及仪器仪表中至关重要。信号处理电路需要先进的分析和设计技术,如傅里叶分析、拉普拉斯变换和滤波器设计。

4.4 射频电路设计

射频电路设计研究在射频下工作的电路。它涉及放大器、混频器、滤波器和天线的设计和分析。射频电路用于通信系统、雷达和无线设备。

4.5 电力电子技术

电力电子技术研究转换和控制电力的电路。它涉及电源、电机驱动器和电源转换器的设计和分析。电力电子技术在可再生能源系统、电动汽车和工业自动化中至关重要。

总之,先进的电子电路分析和设计需要各个领域的专业知识,如非线性电路分析、数字电路分析、信号处理电路、射频电路设计和电力电子技术。通过了解这些领域,工程师可以设计和分析满足现代电子设备要求的复杂电子电路。

Text A

Basic Principles of Control

1. The Principle of Closed-loop Control

The principle of closed-loop control consists in the value to be controlled being fed back from the place of measurement via the controller and its settings' facility into the controlled system (see Figure 7-1).[1] The feedback process makes this so-called controlled variable more independent of external and internal disturbance variables, and is the factor which enables a desired value, the setpoint, to be adhered to in the first place.[2] As the manipulated variable output by the controller influences the controlled variable, the so-called control loop is duly closed.

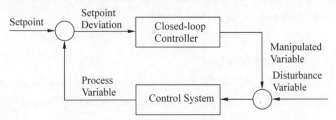

Figure 7-1   The principle of closed-loop control

Technical systems process several kinds of controlled variables such as current, voltage, temperature, pressure, flow, speed of rotation, angle of rotation, chemical concentration and many more. Disturbance variables are also of a physical nature.

These control loop terms can be easily explained using the familiar example of room temperature control by means of a radiator thermostat: the requirement is to keep the room temperature at 22°C. This temperature is set by means of the rotatable knob (= setpoint). The temperature (= controlled variable) is measured by a sensor. The deviation between the room temperature and the setpoint is then measured by the built-in controller, often a bimetal spring, (= the control deviation), and is then used to open or close the valve (= the manipulated variable).

What are the disturbance variables? First of all there is the effect of the outside temperature and the sun shining through the windows. The "limited" thermostat can as little foresee these influences as it can the occupant's behaviour in opening a window or holding a party with a lot of guests who cause the room to warm up.[3] However this controller is still able to compensate for the effects of one or more disturbance variables, and to bring the temperature back to the desired level again, albeit with some delay.

2. The Principle of Open-loop Control

Open-loop control is to be found wherever there is no closed-control loop (see Figure 7-2). The biggest disadvantage compared with closed-loop control is that unknown or non-measurable disturbance variables cannot be compensated. Also the behaviour of the system including the effects of disturbance variables which the open-loop control system is able to measure, must be exactly known at all times in order to be able to use the manipulated variable to influence the controlled variable.[4]

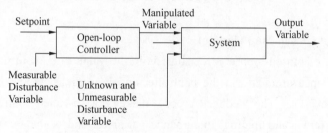

Figure 7-2 The principle of open-loop control

One advantage is that an open-loop control system cannot become instable as there is no feedback—this is a problem of closed-loop control.

3. Classification of Closed-loop Systems

Closed-loop systems are not classified by the physical values to be controlled, but by their behaviour over time. The level in a container can thus be mathematically described in exactly the same way as the voltage of a capacitor.

Behaviour over time can be determined for example by abruptly changing the input value and then observing the output value. Knowledge of the basic laws of physics is often sufficient to estimate this behaviour. Only in relatively few cases is it necessary to calculate it.

The behaviour over time of a closed-loop system is normally characterised by the fact that when the input value is abruptly changed, although the output value immediately begins to change, it reaches its end value with some delay (see Figure 7-3).[5]

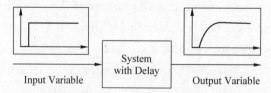

Figure 7-3 Control system with delay

Closed-loop systems are further distinguished by those with and those without self-regulation.

In a system with self-regulation, after the sudden change in the input value, the output value assumes a constant value again after a period of time. Such systems are usually called proportional systems or P systems. Let us take the example of a heating zone: the input value is the electrical heating power, and the output value is the zone temperature. (see Figure 7-4)

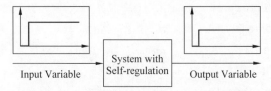

Figure 7-4　Control system with self-regulation

In a system which does not have self-regulation, the output value will rise or fall after the abrupt change in the input value (see Figure 7-5). The output will only remain at a constant value if the input value is at zero. Such systems are usually called integral systems or I systems. An example of this is a level control in a container: the input value is the incoming flow, the output value is the level of the liquid.

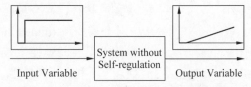

Figure 7-5　Control system without self-regulation

Another important type of system is a system with dead time (see Figure 7-6). In this case the input value appears at the output after the dead time delay. In a technical system the dead time is the result of the distance between setting and measuring locations. Example of a conveyor belt: the input value is the quantity of material at the beginning of the belt, and the output value is the measurement of the amount at the end of the belt. The dead time is calculated as the length of the belt divided by its speed, and it can therefore vary.

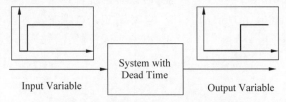

Figure 7-6　Control system with dead time

In all the different systems discussed which are normally found in combination, we are dealing with so-called linear single variable systems because there is only one output value (= the controlled variable) as well as one input value (= the manipulated variable), and the system possesses linear properties.[6]

New Words

principle　　　　　['prɪnsəpl]　　　　*n.* 法则，原则，原理
closed-loop　　　['kləʊzd 'luːp]　　 *n.* 闭环

controller	[kən'trəʊlə]	n. 控制器
facility	[fə'sɪlətɪ]	n. 容易，熟练；便利，敏捷；设备，工具
feedback	['fiːdbæk]	n. 反馈，反应
factor	['fæktə]	n. 因素，要素
duly	['djuːlɪ]	adv. 适当，合适，适度；当然；及时
built-in	[ˌbɪlt 'ɪn]	adj. 内置的，固定的，嵌入的
disturbance	[dɪ'stɜːbəns]	n. 打扰，干扰
manipulate	[mə'nɪpjuleɪt]	vt. （熟练地）操作，使用（机器等）
thermostat	['θɜːməstæt]	n. 自动调温器，温度调节装置
rotatable	[rəʊ'teɪtəbl]	adj. 可旋转的，可转动的
knob	[nɒb]	n. （门、抽屉等的）球形把手，旋钮
sensor	['sensə]	n. 传感器
deviation	[ˌdiːvɪ'eɪʃn]	n. 偏离，偏向，偏差
bimetal	['baɪ'metl]	n. 双金属材料，双金属器件 adj. 双金属的（=bimetallic）
spring	[sprɪŋ]	n. 弹簧，发条；弹性，弹力
disadvantage	[ˌdɪsəd'vɑːntɪdʒ]	n. 缺点，不利，不利条件，劣势
classify	['klæsɪfaɪ]	vt. 分类，分等
behaviour	[bɪ'heɪvjə]	n. 行为，举止，习性
sufficient	[sə'fɪʃnt]	adj. 充分的，足够的
abruptly	[ə'brʌptlɪ]	adv. 突然地，唐突地
relatively	['relətɪvlɪ]	adv. 相关地
characterise	['kærɪktəraɪz]	vt. 表现……特点，具有……特征（=characterize）
estimate	['estɪmeɪt]	vt. & vi. 估计，估价，评估
	['estɪmət]	n. 估计，估价，评估
self-regulation	[self ˌregjʊ'leɪʃn]	n. 自动调节，自调整，自调节
belt	[belt]	n. 带子，传送带，传动带

Phrases

consist in	存在于
be independent of	不依赖……，独立于……
adhere to	坚持
compare with...	与……比较
warm up	变暖，热身
over time	超时，滞后
proportional system	比例系统
dead time	死区

Notes

[1] The principle of closed-loop control consists in the value to be controlled being fed back from the place of measurement via the controller and its settings' facility into the controlled system.

本句是个简单句，谓语动词是 consists in（存在于……），宾语由两部分组成：the value to be controlled being fed back from the place of measurement via the controller 及 its settings' facility into the controlled system。这两个宾语都是由名词短语组成，分别表示两个动作：被控制的值反馈回来（value being fed back）及设定的值迅速进入控制系统（settings' facility into the controlled system）。

本句意为：闭环控制的原理是通过控制器测量反馈回来的控制值，并与设定值一起输入控制系统。

[2] The feedback process makes this so-called controlled variable more independent of external and internal disturbance variables, and is the factor which enables a desired value, the setpoint, to be adhered to in the first place.

本句是由 and 连接的两个并列句，它们有一个共同的主语 feedback process。第一个句子的谓语是 makes，this so-called controlled variable 是宾语，more independent of external and internal disturbance variables 是宾语补足语。第二个句子是一个系表结构，is the factor which enables a desired value, the setpoint, to be adhered to in the first place 是表语，在该表语中，which enables a desired value, the setpoint, to be adhered to in the first place 是定语从句，修饰和限定 the factor。

本句意为：反馈过程使得所谓的控制变量更加独立于外部和内部的干扰变量，它也是使得期望值即设定值始终被保持在起始位置的重要因素。

[3] The "limited" thermostat can as little foresee these influences as it can the occupant's behaviour in opening a window or holding a party with a lot of guests who cause the room to warm up.

本句包含一个比较结构 as little…as…。little 有否定含义，否定的是后面的 foresee（预见）。it can 后省略了动词 foresee。这个结构比较的是不能预见的程度。who cause the room to warm up 表面上是 guests 的定语，其实表达的是一种结果：聚会的客人能使屋内温度上升。

本句意为："有限"自动温控器不能预知这些影响，就如同不能预知人们开窗和举办聚会的行为（聚会的客人能使屋内温度上升）。

[4] Also the behaviour of the system including the effects of disturbance variables which the open-loop control system is able to measure, must be exactly known at all times in order to be able to use the manipulated variable to influence the controlled variable.

本句的主语是 the behaviour of the system，谓语是 must be exactly known，in order to be able to use the manipulated variable to influence the controlled variable 是目的状语。including the effects of disturbance variables 修饰 the behaviour of the system，后面由 which 引导的定语从句修饰 variables。at all times 的意思是"总是；随时；永远"。

本句意为：此外，必须始终准确地了解系统的行为，包括开环控制系统能够测量的干扰变量的影响，以便能够使用操作变量来影响控制变量。

[5] The behaviour over time of a closed-loop system is normally characterised by the fact that when the input value is abruptly changed, although the output value immediately begins to change, it reaches its end value with some delay.

本句中，主语是 The behaviour over time of a closed-loop system，谓语是 is characterised by，后续为具体"特点"。that when the input value is abruptly changed, although the output value immediately begins to change, it reaches its end value with some delay 是一个同位语从句，对 the fact 进行具体说明。在该同位语从句中，when 引导了一个时间状语从句，although 引导了一个让步状语从句，it reaches its end value with some delay 是主句。over time 本意为"随着时间的过去"，behaviour over time 意思为"滞后动作"。

本句意为：闭环控制系统的滞后动作的特点通常为，当输入值突然变化时，尽管输出值立即开始变化，但达到最终的值需要一些延时。

[6] In all the different systems discussed which are normally found in combination, we are dealing with so-called linear single variable systems because there is only one output value (= the controlled variable) as well as one input value (= the manipulated variable), and the system possesses linear properties.

本句中，which are normally found in combination 是一个定语从句，修饰和限定 all the different systems。we are dealing with so-called linear single variable systems 是主句，because 引导了一个表示原因的状语从句，它由 and 连接的两个并列句组成。In all the different systems 限定 we are dealing with 的范围。

本句意为：这里讨论的不同的系统通常以组合形式出现。我们讨论的是所谓的线性单变量系统，因为它只有一个输出值（控制变量），也只有一个输入值（操作变量），并且系统具有线性属性。

Exercises

【**Ex.1**】根据课文内容，回答以下问题。

1. What is the principle of closed-loop control?

2. What is the biggest disadvantage of open-loop control?

3. What is the advantage of open-loop control?

4. How to classify closed-loop systems?

5. What is the characteristic of proportional system?

【Ex.2】把下面的单词填写到合适的空格中。

| A | systems | B | connection | C | applications | D | environments | E | magnetic |
| F | automation | G | electronic | H | electronics | I | functionality | J | open |

What Is a Relay?

A relay is an electrical switch that controls the flow of current in an electrical circuit. It acts as a mediator between two circuits, allowing them to interact without direct physical ___1___. This makes relays an essential component in various ___2___ devices and systems.

Now, let's explore some key aspects of relays:

Function: relays function by using an electrical coil to create a ___3___ field, which pulls or releases mechanical contacts. When the coil is energized, the contacts close, allowing current to flow. When the coil is de-energized, the contacts ___4___, interrupting the current flow.

Applications: relays find applications in numerous industries and ___5___. Some common uses include controlling motors, lighting systems, HVAC (Heating, Ventilation, and Air Conditioning) equipment, and safety systems. They are also used in complex industrial ___6___ processes, automotive systems, telecommunications, and much more.

Types of Relays: there are several types of relays available, each designed for specific ___7___. These include electromechanical relays, solid-state relays, thermal relays, time-delay relays, and reed relays. Each type offers unique features and advantages, catering to different requirements and ___8___.

In conclusion, relays play a crucial role in the world of ___9___ by facilitating the control and interaction between different electrical circuits. Their applications are vast and varied, spanning across industries and systems. Understanding the basic ___10___ and types of relays can help you comprehend their significance and enhance your knowledge of electrical systems.

【Ex.3】把下列句子翻译成中文。

1. Process control is a term applied to the control of variable in a manufacturing process.

2. Much current work in process control involves extending the use of the digital computer to provide direct digital control (DDC) of the variables.

3. Numerical control (NC) is a system that uses predetermined instructions called a program to control a sequence of such operation.

4. In direct digital control the computer calculates the values of the manipulated variables directly from the values of the set points and the measurements of the process variables.

5. Aircraft flight control has been proven to be one of the most complex control applications due to the wide range of system parameters and the interaction between controls.

6. A linear device is one where output is directly proportional to its input(s) and any dynamic function.

7. A pneumatic controller ceases to operate linearly when its output falls to zero or reaches full supply pressure.

8. By automatic, it generally means that the system is usually capably of adapting to a variety of operating conditions and is able to respond to a class of inputs satisfactorily.

9. As the motor does not demands very high accuracy, the open-loop control is used.

10. A feedback loop is the part of a system in which some portion (or all) of the system's output is used as input for future operations.

【Ex.4】把下列短文翻译成中文。

Open Loop Control Systems and Closed Loop Control Systems

1. Open Loop Control Systems

The simplest type of control system is the open loop control system. In an open loop control system, the controller makes decisions based on a predetermined set of inputs and assumptions, without considering feedback about the actual state of the system. As a result, open loop control systems are typically less accurate and unable to account for external disturbances that may affect system performance. Open loop control systems are often used in situations where accuracy is not critical, such as in simple machines or appliances.

2. Closed Loop Control Systems

Closed loop control systems take into account feedback about the actual state of the system and adjust accordingly. This way, closed loop control systems are able to maintain stable system performance even when external disturbances are present. Closed loop control systems are better suited for complex systems where accuracy and stability are critical, such as in industrial processes or aerospace applications.

One example of a closed loop control system is the cruise control system in a car. The controller receives feedback from the speed sensor and adjusts the throttle accordingly to maintain a constant speed. Another example is the heating and cooling system in a building. The controller receives feedback from temperature sensors and adjusts the HVAC (Heating, Ventilation and Air Conditioning) system to maintain a comfortable temperature.

【Ex.5】通过 Internet 查找资料，借助电子词典、辅助翻译软件及 AI 工具，提交一篇简明技术报告，内容包括开环控制和闭环控制的基本知识和典型电路（附各种典型控制系统的电路框图），并附上收集资料的网址。通过 E-mail 发送给老师，或按照教学要求在网上课堂提交。

Text B

Digital Control Systems

With the advent of microprocessors and the low cost of computer hardware, greater flexibility and in general better control can be obtained by using a digital computer as the controller.

The structure of a digital computer controlled system is as follows (see Figure 7-7).

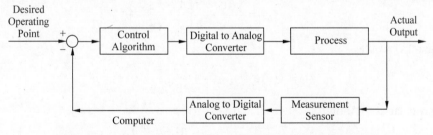

Figure 7-7　The structure of a digital computer controlled system

Now we have part of the system as discrete (digital section) and the other as continuous (process and sensors).

Convert digital part to continuous and do mathematics in s or t domain—difficult, gives transcendental equations to solve.

Convert to discrete and do mathematics in z domain— this is easier but care must be taken. This approach gives no information about signals between samples, and the process being continuous could produce large variations between samples.

In week 1st a simple approximation to obtain the discrete model was presented. It was
$$\frac{dy}{dt} \approx \frac{y(k+1)-y(k)}{\Delta t}$$
and for a 1st order system of the form
$$\frac{dy}{dt} = -ay + u \tag{1}$$
resulted in the difference equation:
$$y(k+1) = (1-a\Delta t)y(k) + \Delta t u(k) \tag{2}$$

Note Δt is the sampling interval, and $u(k)$ is assumed constant during this sampling interval. In the exercises for weeks 1st and 2nd it was shown that even for a constant input, e.g. a unit step, equation(2) y values did not correspond to the y values from equation(1) at the sample points. This was because of the approximation used for the derivative.

Is there a better conversion to discrete form that will be the same at the sampling instances for an input $u(t)$ that is constant during the sampling period? This constant input would be the result of the output from a D/A converter in a digital computer, and hence the following analysis is very relevant to computer controlled systems. Taking L.T.s of equation(1) gives
$$sY(s) - y(0) = -aY(s) + U(s)$$
$$Y(s) = \frac{y(0)}{s+a} + \frac{U(s)}{s+a}$$

Consider the first sample period

where $u(0)$ is constant and hence could be considered as a step of size A. i.e. $U(s) = A/s$
$$Y(s) = \frac{y(0)}{s+a} + \frac{A}{s(s+a)}$$
$$= \frac{y(0)}{s+a} + \frac{k_1}{s} + \frac{k_2}{s+a}$$
$$= \frac{y(0)}{s+a} + \frac{A}{a}\left(\frac{1}{s} - \frac{1}{s+a}\right)$$

Inverse L.T.s gives
$$y(t) = y(0)e^{-at} + \frac{A(1-e^{-at})}{a}, \quad t \geq 0$$

In particular at $t = \Delta t$ or T
$$y(\Delta t) = y(0)e^{-a\Delta t} + \frac{A(1-e^{-a\Delta t})}{a} \tag{3}$$

Noting that $u(t)$ is constant between samples and that the above derivation would be equally valid for any time period, then equation(3) can be generalized to
$$y(k+1) = e^{-a\Delta t} y(k) + \frac{(1-e^{-a\Delta t})u(k)}{a} \tag{4}$$

Taking z-transforms gives
$$zY(z) = e^{-a\Delta t} Y(z) + \frac{(1-e^{-a\Delta t})U(z)}{a}$$

giving

$$\frac{Y(z)}{U(z)} = \frac{(1-e^{-a\Delta t})}{a(z-e^{-a\Delta t})} \quad (5)$$

Equation(4) is often written as

$$y(k+1) = e^{-aT} y(k) + \frac{(1-e^{-aT})u(k)}{a} \quad (6)$$

Where T is the sampling period.

1. Example —Car Velocity Model

Continuous model

$$\frac{v(s)}{m(s)} = \frac{2/15}{s+1/15}$$

For $m(s) = 1/s$

$$v(s) = \frac{2/15}{s(s+1/15)} = \frac{k_1}{s} + \frac{k_2}{s+1/15}$$

Inverse L.T.s gives

$$v(t) = \frac{2}{15}\left[\frac{15}{1}(1-e^{-t/15})\right], \quad t \geq 0 \quad (7)$$

$$= 2(1-e^{-t/15}), \quad t \geq 0$$

Discrete accurate model

$$\frac{v(z)}{m(z)} = \frac{\frac{2}{15} \times \frac{15}{1}(1-e^{-t/15})}{z-e^{-t/15}} = \frac{2(1-e^{-t/15})}{z-e^{-t/15}}$$

If $t = 1$ s

$$v(k+1) = 0.936v(k) + 0.128m(k)$$

Discrete approximate model

$$\frac{v(k+1) - v(k)}{1} = -\frac{1}{15}v(k) + \frac{2}{15}m(k)$$

giving

$$v(k+1) = 0.933v(k) + 0.133m(k)$$

Students should simulate the three responses and verify that the discrete accurate conversion response lies exactly on the continuous response and that an error exists between the discrete approximate conversion response and the other two responses.

The above work outlined a more accurate method for finding the discrete model of a continuous system. This approach was only valid for systems with inputs that were constant during the sampling period, and hence is particularly suitable for digital control systems. The method was applied to a first order system, but can readily be extended to higher order systems.

2. PID Controller

One of the most successful controllers in industry is known as a PID (Proportional, Integral,

Derivative) controller. In the continuous time domain it has the form

$$u(t) = K_p e(t) + K_i \int_0^t e(t)\,dt + K_d \frac{de(t)}{dt} \tag{8}$$

where $e(t)$ is the error signal, and $u(t)$ the control input to the process. In the s-domain

$$U(s) = \left(K_p + \frac{K_i}{s} + K_d s\right) E(s) \tag{9}$$

Note the integral term $K_i \int_0^t e(t)\,dt$ effectively takes into account past errors and uses these to calculate the control $u(t)$ at the present time t. To implement this in a digital computer a discrete approximation is used, and equation(8) becomes

$$u(k) = K_p (e)k + K_i \sum_{j=1}^k e(j)\Delta t + K_d \frac{[e(k) - e(k-1)]}{\Delta t} \tag{10}$$

Note the term $\sum_{j=1}^k e(j)$ does not have to be calculated for all k since a running sum can be made and updated when the new error is calculated. In the z-domain

$$U(z) = \left(K_p + K_i \frac{T_z}{z-1} + \frac{K_d}{T} \frac{(z-1)}{z}\right) E(z) \tag{11}$$

New Words

advent	['ædvent]		*n.* （尤指不寻常的人或事）出现，到来
hardware	['hɑːdweə]		*n.* （计算机的）硬件，（电子仪器的）部件
structure	['strʌktʃə]		*n.* 结构，构造
domain	[də'meɪn]		*n.* 范围，领域
transcendental	[ˌtrænsen'dentl]		*adj.* 先验的，超越的，超出人类经验的
approach	[ə'prəʊtʃ]		*n.* 方法
			vt. 接近，动手处理
			vi. 靠近
sample	['sɑːmpl]		*n.* 标本，样品，例子
			vt. 取样，采样，抽取……的样品
approximation	[əˌprɒksɪ'meɪʃn]		*n.* 近似值，接近，走近
correspond	[ˌkɒrə'spɒnd]		*vi.* 符合，协调，相当，相应
derivative	[dɪ'rɪvətɪv]		*n.* 导数，微商
converter	[kən'vɜːtə]		*n.* 转换器，变换器
relevant	['reləvənt]		*adj.* 有关的，相应的
hence	[hens]		*adv.* 因此，从此
generalize	['dʒenrəlaɪz]		*vt.* 归纳，概括；推广，普及
simulate	['sɪmjuleɪt]		*vt.* 模拟，模仿
accurate	['ækjərət]		*adj.* 准确的，精确的
exactly	[ɪg'zæktlɪ]		*adv.* 精确地，确切地
valid	['vælɪd]		*adj.* 有效的，有根据的，正确的

suitable	[ˈsuːtəbl]	*adj.* 适当的，相配的
transform	[trænsˈfɔːm]	*vt.* 转换，改变，改造，使……变形
		vi. 改变，转化，变换
		n. 变换（式）
velocity	[vəˈlɒsətɪ]	*n.* 速度，速率；迅速；周转率
outline	[ˈaʊtlaɪn]	*n.* 大纲；轮廓，略图，外形；要点，概要
		vt. 描画轮廓，略述
signal	[ˈsɪɡnəl]	*n.* 信号
		adj. 信号的
		vt. & vi. 发信号，用信号通知
integral	[ˈɪntɪɡrəl]	*adj.* 积分的
		n. 积分
implement	[ˈɪmplɪment]	*vt.* 贯彻，实现
		vt. & vi. 执行
	[ˈɪmplɪmənt]	*n.* 工具，器具

Phrases

result in	导致
convert …to …	把……转换为……
sampling period	采样周期
sampling interval	采样间隔

Abbreviations

| D/A(Digit/Analogy) | 数字/模拟 |
| PID (Proportional, Integral, Differential) | 比例-积分-微分（控制器） |

Exercises

【Ex.6】根据文章所提供的信息判断正误。

1. The price of hardware of computer is increasingly high.
2. A digital computer controlled system consist in control algorithm, digital to analog converter, process, analog digital converter and measurement sensor.
3. A digital computer controlled system is only designed for discrete variable.
4. In a digital computer controlled system , digital signal must be converted to analog signal.
5. According to the equation:

$$y(k+1) = (1 - a\Delta t)y(k) + \Delta t u(k)$$

Δt is the sampling interval.

6. According to equation(1) and equation(2), if the input does not change, the output will keep constant.

7. The equation as the following:
$$y(\Delta t) = y(0)e^{-a\Delta t} + \frac{A(1-e^{-a\Delta t})}{a}$$
only applies in special time period.

8. According to the passage, the car velocity model is
$$\frac{v(s)}{m(s)} = \frac{2/15}{s+1/15}$$

9. PID controller is a very useful controller in industry.

10. Past errors can be used to calculate the control $u(t)$ at the present time t in a PID controller.

科技英语翻译知识

被动语态的译法

英语中被动语态的使用范围很广，主动者经常被省去，科技英语中更是如此。因为科技文献侧重叙事和推理，强调的是作者的观点和发明内容，而不是作者本人。汉语里很少使用被动语态，因此，在翻译时，应尽可能将英语的被动句译成汉语的主动句。视具体情况，也可保留被动句。

1. 译成主动句

(1) It is clear that a body can be charged under certain condition.

很显然，在一定条件下物体能够带电。

（原文中的主语 body 在译文中保留为主语。）

(2) Solution to the problem was ultimately found.

这个问题的解决办法终于找到了。

（原文中的 solution 仍然作译文中的主语，不必译成"被找到"。）

(3) If one of more electrons be removed, the atom is said to be positively charged.

如果原子失去一个或多个电子，就说该原子带正电荷。

（原文中的主语 electrons 在译文中作宾语。）

(4) Three-phase current should be used for large motors.

大型电动机应当使用三相电流。

（原文中的主语 three-phase current 译作宾语，而原宾语 large motors 在译文中作主语。）

(5) An emf will be generated in the coil.

线圈中会产生电动势。

（地点状语 in the coil 译成主语。）

(6) Any desired voltage may be obtained by means of a transformer.

借助变压器可获得任何需要的电压。
（方式状语 by means of transformer 提前，译成无主句。）

(7) The resistance can be tested to determine its voltage and current.
只要知道电压和电流，就能测定电阻。
（原文中的主语 the resistance 在译文中做宾语，译成无主句。）

(8) When these charges move along a wire, that wire is said to carry an electric current.
当这些电荷沿导线运动时，就说该导线带电。
（主语 wire 在汉语句中，与后面的 carry an electric current 构成分句。）

有 it 引导的习惯句型中的被动语态，常有固定的译法：

It is hoped that…希望
It is reported that…据报道
It is said that…据说
It can be foreseen that…可以预料
It cannot be denied that…无可否认
It has been proved that…已经证明
It is arranged that…已经商定，已经准备
It is alleged that…据称
It is announced that…普遍认为
It is believed that…大家相信
It is demonstrated that…据事实
It is accepted that…可接受的是

2．译成汉语被动句

译成汉语被动句时，除了用"被"字外，还可使用"受""由""为……所……"等。

(1) Current will not flow continually, since the circuit is broken by the insulating material.
电流不能继续流动，因为电路被绝缘材料隔断了。

(2) Every charged object is surrounded by an electric field.
每个带电体都为电场所包围。

(3) The magnetic field is produced by an electric current.
磁场由电流产生。

(4) Resistivity is affected by temperature, moisture and structural defects.
电阻率受温度、湿度和结构缺陷的影响。

(5) Current flow is measured in amperes（A）and is represented by the letter I.
电流以安培（A）为单位来量度，并用字母 I 来表示。

(6) Our streets and houses are lighted by electricity.
我们的街道和房屋靠电来照明。

(7) The values of current and voltage required can be obtained from the volt-ampere curve or from an actual circuit.
所需电流和电压的值可以从伏安曲线或实际电路获得。

(8) A graph is often more easily understood than a law.

图比定律更容易为人所理解。

Reading Material

阅读下列文章。

Text	Note
Internet of Things	
The internet of things, or IoT, is a network of interrelated devices that connect and exchange data with other IoT devices and the cloud. IoT devices are typically embedded with technology such as sensors and software and can include mechanical[1] and digital machines and consumer objects.	[1] *adj.* 机械的
With IoT, data is transferable[2] over a network without requiring human-to-human or human-to-computer interactions.	[2] *adj.* 可传输的
1. How Does IoT work?	
An IoT ecosystem[3] consists of web-enabled smart devices that use embedded systems — such as processors, sensors and communication hardware — to collect, send and act on data they acquire from their environments. IoT devices share the sensor data they collect by connecting to an IoT gateway[4], which acts as a central hub where IoT devices can send data. Before the data is shared, it can also be sent to an edge device where that data is analyzed locally. Analyzing data locally reduces the volume of data sent to the cloud, which minimizes bandwidth[5] consumption.	[3] *n.* 生态系统 [4] *n.* 网关 [5] *n.* 带宽
Sometimes, these devices communicate with other related devices and act on the information they get from one another. The devices do most of the work without human intervention, although people can interact with the devices — for example, to set them up, give them instructions or access the data.	
The connectivity[6], networking and communication protocols used with these web-enabled devices largely depend on the specific IoT applications deployed.	[6] *n.* 连通性
IoT can also use artificial intelligence and machine learning to aid in making data collection processes easier and more dynamic.	
2. Why Is IoT important?	
IoT helps people live and work smarter. Consumers, for example,	

can use IoT-embedded devices — such as cars, smartwatches or thermostats[7] — to improve their lives. For example, when a person arrives home, their car could communicate with the garage[8] to open the door; their thermostat could adjust to a preset temperature; and their lighting could be set to a lower intensity and color.

In addition to offering smart devices to automate homes, IoT is essential to business. It provides organizations with a real-time look into how their systems really work, delivering insights into everything from the performance of machines to supply chain[9] and logistics[10] operations.

IoT enables machines to complete tedious tasks without human intervention. Companies can automate processes, reduce labor costs, cut down on waste and improve service delivery. IoT helps make it less expensive to manufacture and deliver goods, and offers transparency[11] into customer transactions.

3. What Are the Benefits of IoT to Organizations?

IoT offers several benefits to organizations. Some benefits are industry-specific and some are applicable across multiple industries. Common benefits for businesses include the following:

(1) Monitoring overall business processes.

(2) Improving the customer experience[12].

(3) Saving time and money.

(4) Enhancing employee productivity.

(5) Providing integration and adaptable business models.

(6) Enabling better business decisions.

(7) Generating more revenue.

Generally, IoT is most abundant in manufacturing, transportation and utility organizations that use sensors and other IoT devices; however, it also has use cases for organizations within the agriculture, infrastructure and home automation industries, leading some organizations toward digital transformation.

4. What Are the Pros and Cons of IoT?

Some of the advantages of IoT include the following:

(1) Enabling access to information from anywhere at any time on any device.

(2) Improving communication between connected electronic devices.

---

[7] *n.* 恒温（调节）器

[8] *n.* 车库

[9] supply chain: 供应链

[10] *n.* 物流

[11] *n.* 透明度，透明性

[12] customer experience: 客户体验

(3) Enabling the transfer of data packets[13] over a connected network, which can save time and money.

(4) Collecting large amounts of data from multiple devices, aiding both users and manufacturers.

(5) Analyzing data at the edge, reducing the amount of data that needs to be sent to the cloud.

(6) Automating tasks to improve the quality of a business's services and reducing the need for human intervention[14].

(7) Enabling healthcare patients to be cared for continually and more effectively.

Some disadvantages of IoT include the following:

(1) Increasing the attack surface as the number of connected devices grows. As more information is shared between devices, the potential for a hacker to steal confidential[15] information increases.

(2) Making device management challenging as the number of IoT devices increases. Organizations might eventually have to deal with a massive number of IoT devices, and collecting and managing the data from all those devices could be challenging.

(3) Having the potential to corrupt[16] other connected devices if there's a bug in the system.

(4) Increasing compatibility issues between devices, as there's no international standard of compatibility for IoT. This makes it difficult for devices from different manufacturers to communicate with each other.

5. IoT Standards

Some examples of IoT standards include the following:

(1) IPv6 over low power wireless personal area networks (6LoWPAN). It is an open standard defined by the Internet engineering task force (IETF). This standard enables any low-power radio to communicate to the internet, including 804.15.4, bluetooth[17] low energy and Z-Wave for home automation. In addition to home automation, this standard is also used in industrial monitoring and agriculture.

(2) ZigBee. ZigBee is a low-power, low-data rate wireless network used mainly in home and industrial settings. It is based on the IEEE 802.15.4 standard. The ZigBee Alliance created Dotdot, the universal language for IoT that enables smart objects to work securely on any network and understand each other.

---

[13] data packet: 数据包

[14] *n.* 介入，干预

[15] *adj.* 秘密的，机密的

[16] *v.* 损坏

[17] *n.* 蓝牙（一种无线传输技术）

(3) Data distribution service (DDS). It was developed by the Object Management Group and is an industrial IoT (IIoT) standard for real-time, scalable and high-performance machine-to-machine[18] (M2M) communication.

IoT standards often use specific protocols for device communication. A chosen protocol dictates how IoT device data is transmitted and received. Some example IoT protocols include the following:

(1) Constrained application protocol (CoAP). It is a protocol designed by the IETF that specifies how low-power, compute-constrained devices can operate in IoT.

(2) Advanced message queuing protocol (AMQP). It is an open source published standard for asynchronous messaging by wire. AMQP enables encrypted[19] and interoperable[20] messaging between organizations and applications. The protocol is used in client-server messaging and in IoT device management.

(3) Long range wide area network (LoRaWAN). This protocol for WANs[21] is designed to support huge networks, such as smart cities[22], with millions of low-power devices.

(4) MQ telemetry transport (MQTT). It is a lightweight[23] protocol that's used for control and remote monitoring applications. It's suitable for devices with limited resources.

6. Consumer and Enterprise IoT Applications

There are numerous real-world applications of the internet of things, ranging from consumer IoT and enterprise IoT to manufacturing and IIoT. IoT applications span numerous verticals, including automotive, telecom and energy.

In the consumer segment, for example, smart homes that are equipped with smart thermostats, smart appliances and connected heating, lighting and electronic devices can be controlled remotely via computers and smartphones.

Wearable devices with sensors and software can collect and analyze user data and send messages to other technologies about the users with the aim of making their lives easier and more comfortable. Wearable devices[24] are also used for public safety — for example, by improving first responders' response times during emergencies by providing optimized routes to a location or by tracking construction workers' or firefighters' vital signs at life-threatening sites.

[18] 机器对机器

[19] *v.* 加密
[20] *adj.* 可互操作的

[21] Wide Area Network: 广域网
[22] smart city: 智慧城市
[23] *adj.* 轻量的

[24] wearable device: 穿戴式设备

In healthcare, IoT offers providers the ability to monitor patients more closely using an analysis of the data that's generated. Hospitals often use IoT systems to complete tasks such as inventory management for both pharmaceuticals and medical instruments.

Smart buildings can, for instance, reduce energy costs using sensors that detect how many occupants are in a room. The temperature can adjust automatically — for example, turning the air conditioner[25] on if sensors detect a conference room is full or turning the heat down if everyone in the office has gone home.

[25] air conditioner: 空调机；空调设备

In agriculture, IoT-based smart farming systems can help monitor light, temperature, humidity and soil moisture of crop fields using connected sensors. IoT is also instrumental in automating irrigation systems.

In a smart city, IoT sensors and deployments, such as smart streetlights and smart meters, can help alleviate traffic, conserve energy, monitor and address environmental concerns and improve sanitation.

7. IoT Security and Privacy Issues

IoT connects billions of devices to the internet and involves the use of billions of data points, all of which must be secured. Due to its expanded attack surface, IoT security and IoT privacy are cited as major concerns.

Because IoT devices are closely connected, a hacker can exploit one vulnerability to manipulate[26] all the data, rendering it unusable. Manufacturers who don't update their devices regularly —or at all— leave them vulnerable to cybercriminals[27]. Additionally, connected devices often ask users to input their personal information, including names, ages, addresses, phone numbers and even social media accounts — information that's invaluable to hackers.

[26] *vt.* 操作，处理，操纵

[27] *n.* 网络犯罪分子

Hackers aren't the only threat to IoT; privacy is another major concern. For example, companies that make and distribute consumer IoT devices could use those devices to obtain[28] and sell user personal data.

[28] *vt.* 获得，得到

参 考 译 文

控制的基本原理

1. 闭环控制原理

闭环控制的原理是通过控制器测量反馈回来的控制值，并与设定值一起输入控制系统（见图 7-1）。反馈过程使得所谓的控制变量更加独立于外部和内部的干扰变量，它也是使得期望值即设定值始终被保持在起始位置的重要因素。当由控制器输出的操作变量影响控制变量时，所谓的控制环会适时地关闭。

（图略）

技术系统可以处理多种控制变量，如电流、电压、温度、压力、流量、旋转速度、转动角、化学浓度等。干扰变量也有物理属性。

这些控制环的术语可以用大家熟悉的利用冷却控制器进行室温控制的例子简单地解释：为了将室温保持在 22 ℃，温度可以通过可旋转的旋钮进行设定（设定值）。温度（控制变量）是由传感器测量的。室内温度和设定点之间的偏差由内置控制器（通常是双金属弹簧）测量（控制偏差），然后再用于打开或关闭阀门（操作变量）。

什么是干扰变量？首先是室外温度和透过窗户照射进来的阳光的影响。"有限"自动温控器不能预知这些影响，就如同不能预知人们开窗和举办聚会的行为（聚会的客人能使屋内温度上升）。然而控制器依然可以补偿一种或多种干扰变量带来的影响，尽管有些延时，但可使温度重新回到期望的水平。

2. 开环控制原理

不形成闭环的控制就是开环控制（见图 7-2）。与闭环控制相比，开环控制的最大不足是，对于未知或不能测量的干扰变量无法进行补偿。此外，必须始终准确地了解系统的行为，包括开环控制系统能够测量的干扰变量的影响，以便能够使用操作变量来影响控制变量。

（图略）

开环控制系统的一个优点是，由于没有反馈而使系统稳定——这正是闭环系统的问题所在。

3. 闭环控制系统的分类

闭环系统不是按照被控对象的物理值，而是按照其滞后动作来分类的。因此，容器中的液位可以用与电容器的电压完全相同的方式进行数学描述。

例如，可以通过突然改变输入值然后观察输出值来确定滞后动作。通常根据基本的物理定律的知识就可以充分估计这种动作。只有在相对较少的情况下才需要计算。

闭环控制系统的滞后动作的特点通常为，当输入值突然变化时，尽管输出值立即就开始变化，但达到最终的值需要一些延时（见图 7-3）。

（图略）

闭环系统可以根据是否有自调进一步区分。

在有自调的系统中，当输入值突然变化后，经过一段时间后输出又表现为一个恒定值（见图 7-4）。这种系统常称为比例系统或 P 系统。举一个加热环的例子：输入值是加热电能，输出值是环的温度。

（图略）

在没有自调的系统中，当输入值突然变化后，输出值也会上升或下降（见图 7-5）。只有当输入在零点时输出才保持恒定。这种系统一般称为积分系统或 I 系统。容器的液位控制就是一个例子：注入的流量是输入值，液体的液位是输出值。

（图略）

另一种重要的系统类型是有死区的系统（见图 7-6）。这种情况下输入的改变只有经过死区时间延迟后才能在输出上反映出来。在技术系统中，设定位置与测量位置之间的距离产生了死区时间。例如传送带：输入值是传送带起始处的物料量，输出值是传送带末端的物料量的测量值。死区时间的计算方式为传送带的长度除以其速度，因此它可能会有所不同。

（图略）

这里讨论的不同的系统通常以组合形式出现。我们讨论的是所谓的线性单变量系统，因为它只有一个输出值（控制变量），也只有一个输入值（操作变量），并且系统具有线性属性。

Text A

Electronic Design Automation

In the simplest terms, electronic design automation (EDA) is a process that helps designers create and test electronic circuits. It can be used for a variety of tasks, including schematic capture, simulation, prototyping, and PCB layout.

1. What Is Electronic Design Automation?

Electronic design automation or EDA is the category of software tools used by engineers to design and test electronic systems. These tools automate various tasks in the process of electronic design, such as circuit simulation, chip design/layout, and circuit verification.

Electronic design automation tools are used by engineers to create designs for integrated circuits (ICs), printed circuit boards (PCBs), and field programmable gate arrays (FPGAs). They are also used to create system on chip (SoC) designs and embedded system designs. In short, if you see an electronic device with the elements mentioned above, it was likely designed using electronic design automation tools.

The most common electronic design automation tools can be roughly divided into three categories:

(1) Design entry: these are used to create the schematics or HDL code that describes the design.

(2) Simulation: these are used to verify the functionality of the design.

(3) Physical design: these are used to transform the logical design into a physical layout.

EDA tools are used throughout the design process, from creating the initial schematic diagram of a circuit, to simulating its behaviour, to verifying that it meets all the necessary design requirements. In recent years, EDA tools have become increasingly powerful and sophisticated, due to the ever-growing complexity of electronic devices.

Designers now rely heavily on EDA tools to help them verify the correctness of their designs and meet manufacturing deadlines. Without EDA tools, it would be virtually impossible to create today's complex electronic devices in a timely and cost-effective manner. [1]

2. How Does Electronic Design Automation Work?

Broadly speaking, EDA software aids engineers in the design, analysis, and optimization of circuit elements and systems. By automating certain tasks, such as repetitive calculations or the placement and routing of components, EDA tools help engineers save time and increase accuracy. There are many different types of EDA tools available on the market, each of which is designed to tackle a specific stage of the design process. [2]

The three main stages are: component/system level design, circuit level design, and layout level design.

Component/system level design is where engineers decide on the overall architecture of their circuit or system. This includes defining the functionality and interfaces of each component as well as how those components will interact with one another.

Circuit level design is where engineers go into more detail about each individual component. This includes specifying the values of resistors, capacitors, inductors, etc. as well as designing the circuitry that will connect those components.

Layout level design is where engineers take all of the information from the previous two stages and translate it into a physical layout. This is usually done by using computer aided design (CAD) software to create a 2D or 3D representation of the circuit board.

Once all three stages are complete, the final step is to verify that the circuit behaves as expected by running simulations using EDA simulation tools. These simulations allow engineers to test their designs under a variety of conditions to ensure that they meet all specifications.

By automating various tasks throughout each stage in design, from simple repetitive calculations to more complex processes such as placing and routing components, EDA provides valuable assistance that improves accuracy and timeliness in projects.

3. The Difference Between Manual Design, CAD and EDA

Manual design is exactly what it sounds like — product designers create everything by hand without the help of any software or computers. This was the standard method for designing and building electronics until the late 20th century. While it's possible to create high quality products using manual design methods, it's often very time consuming and expensive.

CAD software is computer aided design software that helps product designers create 2D and 3D models of their products. It is often used in conjunction with manual design methods, but it can also be used on its own. Generally speaking, products that are designed using CAD software are easier and less expensive to produce than those that are designed entirely by hand.[3]

EDA software takes things one step further by automating many of the tasks involved in product design. In addition to helping product designers create models of their products, EDA software can also help with tasks such as managing component libraries, simulating electrical circuits, and generating manufacturing documentation. Because EDA automates many of the tasks involved in product design, it can help businesses save a significant amount of time and money.

4. How to Choose an EDA Tool?

EDA tools can help you save time, reduce errors, optimize performance, and enhance creativity. But with so many options available, how do you choose the best EDA tools for your projects? We will explore some of the key factors and features to consider when selecting EDA tools.

4.1 Project scope and complexity

The first thing to consider when choosing EDA tools is the scope and complexity of your project. Depending on the type, size, and level of detail of your design, you may need different EDA tools to handle different aspects of the design process. For example, if you are designing a simple logic circuit, you may only need a schematic editor and a logic simulator. But if you are designing a complex system on chip (SoC), you may need a suite of EDA tools that can handle high-level synthesis, hardware description languages, physical design, verification, and testing.

4.2 Compatibility and integration

Another important factor to consider when choosing EDA tools is their compatibility and integration with other tools and platforms. You want to make sure that your EDA tools can work well with each other, as well as with other software and hardware that you may use for your project. For example, if you are using a specific microcontroller or FPGA, you want to make sure that your EDA tools can support its programming and debugging. Or if you are using a cloud-based platform for collaboration and storage, you want to make sure that your EDA tools can connect and synchronize with it.

4.3 Usability and functionality

A third factor to consider when choosing EDA tools is their usability and functionality. You want to choose EDA tools that are easy to learn, use, and customize as well as offer the features and functions that you need for your project. For example, if you are designing a mixed-signal circuit, you want to choose an EDA tool that can handle both analog and digital components and signals. Or if you are designing a PCB, you want to choose an EDA tool that can offer layout, routing, and manufacturing options.

4.4 Cost and availability

The final factor to consider when choosing EDA tools is their cost and availability. You want to choose EDA tools that fit your budget and are accessible to you. Depending on your project and preferences, you may opt for free or open-source EDA tools, commercial or proprietary EDA tools, or a combination of both. You also want to check the licensing and support options of your EDA tools as well as their updates and upgrades.

5. The Future of Electronic Design Automation

The field of electronic design automation is always evolving. Here are some important trends in electronic design automation that you might need to know about.

5.1　Machine learning for design optimization

In recent years, machine learning has been applied to many different areas of electronic design with promising results. Machine learning can be used to optimize a design for various performance metrics, such as power consumption, speed and area. This is a trend that is sure to continue in the years to come as designers look for ways to get more out of their designs with less effort. [4]

5.2　Cloud-based tools

The cloud has been a game-changer for many industries and electronic design is no exception. Cloud-based electronic design automation tools offer many advantages over traditional desktop-based tools, such as lower costs, easier collaboration and better scalability. This trend is only going to continue as more and more designers flock to the cloud.

5.3　Integrating analogue and digital designs

In the past, analogue and digital designs were created separately and then integrated at the last minute before fabrication. However, this approach is no longer feasible in today's fast-paced world, where time-to-market is critical. Designers are now integrating analogue and digital designs from the beginning to create a more efficient workflow and avoid potential problems down the road.

5.4　Low-power/energy-efficient design methodologies

Power consumption is a major concern for today's designers due to the ever-increasing demand for mobile devices that can operate for long periods of time without being plugged in. As a result, low-power/energy-efficient design methodologies have become increasingly popular in recent years. These methods help designers meet the stringent power requirements of today's devices while still maintaining performance levels that consumers demand.

5.5　Saving time with automated test benches

In order to stay competitive, designers need to be able to get their products to market as quickly as possible without sacrificing quality or performance levels. One way to achieve this goal is by using automated test benches to verify the functionality of a design before it goes into production. By using automated test benches, designers can save valuable time that would otherwise be spent manually testing their designs. They can spend the time on other tasks such as optimization or designing the next generation of products.

The field of electronic design automation is constantly evolving, which means that what's considered cutting edge today may be old news by tomorrow. However, by staying on top of the latest trends, you can stay ahead of the curve and keep your designs running smoothly, no matter what changes come your way.

New Words

prototype	['prəʊtətaɪp]	n. 原型，雏形，蓝本
automate	['ɔːtəmeɪt]	v. 使自动化
verification	[ˌverɪfɪ'keɪʃn]	n. 验证，证明，证实，核实
initial	[ɪ'nɪʃl]	adj. 开始的，最初的
correctness	[kə'rektnəs]	n. 正确性
deadline	['dedlaɪn]	n. 截止时间，最后期限
aid	[eɪd]	n. & v. 辅助，帮助
route	[ruːt]	vt. 给……规定路线
		n. 路径，途径
tackle	['tækl]	vt. 处理
		n. 用具，装备
condition	[kən'dɪʃn]	n. 条件；状况；环境
specification	[ˌspesɪfɪ'keɪʃn]	n. 规范，规格；详述；说明书
timeliness	['taɪmlɪnəs]	n. 时间性，及时方便
enhance	[ɪn'hɑːns]	v. 提高，增强；改进
creativity	[ˌkriːeɪ'tɪvətɪ]	n. 创造性，创造力
project	['prɒdʒekt]	n. 项目，工程；方案，计划
		v. 规划，计划
editor	['edɪtə]	n. 编辑器，编辑程序
simulator	['sɪmjuleɪtə]	n. 模拟器，模拟程序，模拟装置
synthesis	['sɪnθəsɪs]	n. 综合，综合体
compatibility	[kəmˌpætə'bɪlətɪ]	n. 兼容性
platform	['plætfɔːm]	n. 平台
microcontroller	[ˌmaɪkrəʊkən'trəʊlə]	n. 微控制器
debug	[ˌdiː'bʌg]	vt. 调试；排除故障
collaboration	[kəˌlæbə'reɪʃn]	n. 协作，合作
synchronize	['sɪŋkrənaɪz]	vt. 使同步，使同时
usability	[ˌjuːzə'bɪlɪtɪ]	n. 可用性，适用性
customize	['kʌstəmaɪz]	vt. 定制，定做；用户化
availability	[əˌveɪlə'bɪlətɪ]	n. 可用性；有效性
budget	['bʌdʒɪt]	n. 预算
		v. 把……编入预算
proprietary	[prə'praɪətrɪ]	adj. 专有的，专利的
trend	[trend]	n. 趋势，倾向

scalability	[skeɪləˈbɪlɪtɪ]	n. 可扩展性；可伸缩性
fabrication	[ˌfæbrɪˈkeɪʃn]	n. 制造
feasible	[ˈfiːzəbl]	adj. 可行的
fast-paced	[fɑːst peɪst]	adj. 快节奏的
stringent	[ˈstrɪndʒənt]	adj. 严格的；迫切的
competitive	[kəmˈpetətɪv]	adj. 竞争的；有竞争力的
sacrifice	[ˈsækrɪfaɪs]	n. & v. 牺牲；舍弃

Phrases

in process of	在……的过程中
chip design	芯片设计
circuit verification	电路验证
physical design	物理设计
schematic diagram	原理图，示意图
design process	设计过程
decide on ...	就……做出决定
circuit board	电路板
simulation tool	模拟工具
manual design	手动设计，人工设计
2D model	二维模型
3D model	三维模型
in conjunction with	与……协作
component library	元件库
optimize performance	优化性能
a suite of	一系列；一套
cloud-based platform	基于云的平台
mixed-signal circuit	混合信号电路
machine learning	机器学习
power consumption	耗电量；能量消耗
desktop-based tool	基于桌面的工具
flock to ...	成群结队地走向……
plug in	插上（电源的插头）
low-power/energy-efficient design	节能型设计
test bench	测试工作台，试验工作台
cutting edge	尖端，最前沿；优势

Abbreviations

EDA (Electronic Design Automation)	电子设计自动化
IC (Integrated Circuit)	集成电路

PCB (Printed Circuit Board)　　　　　　印制电路板
FPGA (Field Programmable Gate Array)　现场可编程门阵列
SoC (System on Chip)　　　　　　　　　单片系统
HDL (Hardware Description Language)　 硬件描述语言
CAD (Computer Aided Design)　　　　　 计算机辅助设计

Notes

[1] Without EDA tools, it would be virtually impossible to create today's complex electronic devices in a timely and cost-effective manner.

本句中，Without EDA tools 是介词短语，作条件状语。it 是形式主语，真正的主语是不定式短语 to create today's complex electronic devices。in a timely and cost-effective manner 是方式状语。

本句意为：如果没有 EDA 工具，几乎不可能及时且经济高效地创建当今复杂的电子设备。

[2] There are many different types of EDA tools available on the market, each of which is designed to tackle a specific stage of the design process.

本句中，each of which is designed to tackle a specific stage of the design process 是一个非限定性定语从句，对 EDA tools 进行补充说明。

英语中，名词/代词/数词+of+which / whom 可以引导非限定性定语从句。例如：

Light is the fastest thing in the world, the speed of which is 300000 kilometer per second.

There are 30 chairs in the small hall, most of which are new.

This power plant has over five thousand workers and staffs, eighty per cent of whom are men.

He bought many books last Sunday, five of which are on automation design.

本句意为：市场上有许多不同类型的 EDA 工具，每种工具都旨在解决设计过程中特定阶段的问题。

[3] Generally speaking, products that are designed using CAD software are easier and less expensive to produce than those that are designed entirely by hand.

本句中，that are designed using CAD software 是一个定语从句，修饰和限定 products。that are designed entirely by hand 也是一个定语从句，修饰和限定 those。those 指代 products。

本句意为：一般来说，使用 CAD 软件设计的产品比完全手动设计的产品更容易生产且成本更低。

[4] This is a trend that is sure to continue in the years to come as designers look for ways to get more out of their designs with less effort.

本句中，that is sure to continue in the years to come 是一个定语从句，修饰和限定 a trend。as designers look for ways to get more out of their designs with less effort 是一个原因状语从句。

本句意为：由于设计师寻找事半功倍的方法，因此这种趋势在未来几年肯定会持续下去。

Exercises

【Ex.1】 根据课文内容，回答以下问题。

1. What is electronic design automation? What are the three categories that the most common electronic design automation tools can be roughly divided into?

2. What is layout level design? How is it usually done?

3. What is CAD software?

4. How many key factors and features to consider when selecting EDA tools are mentioned in the passage? What are they?

5. What are some important trends in electronic design automation that you might need to know about?

【Ex.2】 把下面的单词填写到合适的空格中。

A	robotics	B	equipment	C	fields	D	panels	E	sensors
F	manufacturing	G	detection	H	medical	I	control	J	weather

What Is a Sensor?

A sensor is a device or component that detects and responds to physical or environmental changes, converting them into measurable signals or data. Sensors are used in various fields, including electronics, engineering, __1__, healthcare, etc. They play a crucial role in collecting real-world data and enabling automation, control systems, and monitoring processes.

There are numerous types of sensors, each designed to measure specific physical quantities or environmental parameters. Here are some common types of __2__ and their functions:

Temperature sensor: measures temperature variations. Used in thermostats, __3__ monitoring, industrial processes, and HVAC systems.

Pressure sensor: detects changes in pressure. Used in automotive systems, medical devices, industrial __4__, and air pressure monitoring.

Proximity sensor: detects the presence or absence of an object without physical contact. Used in touchless interfaces, object __5__, and robotics.

Motion sensor: detects movement or changes in an object's position. Used in security systems, gaming consoles, automatic doors, and motion-activated lighting.

Light sensor: measures light levels or ambient light. Used in photography, automatic lighting controls, solar ___6___, and display brightness adjustment.

Humidity sensor: measures and monitors humidity or moisture levels. Used in weather stations, HVAC systems, industrial processes, and agriculture.

Gas sensor: detects and measures the presence and concentration of gases. Used in environmental monitoring, gas leak detection, indoor air quality ___7___, and automotive emissions.

Accelerometer: measures acceleration forces. Used in smartphones, gaming controllers, navigation systems, and vehicle stability control.

Gyroscope: measures orientation and rotational motion. Used in drones, virtual reality systems, robotics, and navigation devices.

Magnetic sensor: detects magnetic ___8___. Used in compasses, navigation systems, magnetic switches, and position sensing.

PH sensor: measures the acidity or alkalinity of a solution. Used in water quality monitoring, laboratory analysis, and industrial processes.

Optical sensor: detects and measures light intensity or properties. Used in barcode scanners, visual fiber communication, spectrometry, and ___9___ imaging.

Proximity sensor: measures the distance between the sensor and an object. Used in robotics, automated vehicles, and object detection systems.

Sound sensor: detects sound or acoustic signals. Used in audio recording devices, noise monitoring, and speech recognition systems.

Force sensor: measures the force applied to it. Used in load cells, pressure-sensitive touch screens, ___10___, and industrial processes.

These are just a few examples of sensors and their functions. Many more specialized sensors are available, tailored for specific applications across various industries.

【Ex.3】把下列句子翻译成中文。

1. He wanted to use computers to automate the process.

2. This verification must succeed in order for the sample application to work.

3. The correctness of the proposed method and calculation results are proved by testing results.

4. Requirements engineering consists of activities like requirements elicitation, specification and

validation.

5. The project must be completed within a specific time span.

6. The company uses the simulator to test new designs.

7. This helps you save debugging and troubleshooting time.

8. The design and development of software system on a microcontroller system is particularly important.

9. It is very easy to customize an application for different needs.

10. In addition, the software control system has good stability, flexibility and scalability.

【Ex.4】把下列短文翻译成中文。

What Are the Different Types of Machine Learning ?

There are four major types of machine learning. They are supervised learning, unsupervised learning, semi-supervised learning, and reinforcement learning.

With supervised learning, the computer is provided with a labeled set of data that enables it to learn how to do a human task. This is the least complex model, as it attempts to replicate human learning.

With unsupervised learning, the computer is provided with unlabeled data and extracts previously unknown patterns/insights from it.

With semi-supervised learning, the computer is provided with a set of partially labeled data and performs its task using the labeled data to understand the parameters for interpreting the unlabeled data.

With reinforcement learning, the computer observes its environment and uses that data to identify the ideal behavior that will minimize risk and/or maximize reward. This is an iterative approach that requires some kind of reinforcement signal to help the computer better identify its best action.

【Ex.5】通过 Internet 查找资料，借助电子词典、辅助翻译软件及 AI 工具，完成以下技术报告，并附上收集资料的网址。通过 E-mail 发送给老师，或按照教学要求在网上课堂提交。
1. 当前世界上有哪些最主要的 EDA 软件及简述各个软件的特点。
2. 选择一种 EDA 软件，简述其主要功能。

Text B

SPICE

SPICE (Simulation Program with Integrated Circuit Emphasis) is a general-purpose, open source analog electronic circuit simulator. It is a program used in integrated circuit and board-level design to check the integrity of circuit designs and to predict circuit behavior.

1. Introduction

Unlike board-level designs composed of discrete parts, it is not practical to breadboard integrated circuits before manufacture. Further, the high costs of photolithographic masks and other manufacturing prerequisites make it essential to design the circuit to be as close to perfect as possible before the integrated circuit is first built. Simulating the circuit with SPICE is the industry-standard way to verify circuit operation at the transistor level before committing to manufacturing an integrated circuit.

Board-level circuit designs can often be breadboarded for testing. Even with a breadboard, some circuit properties may not be accurate compared to the final printed wiring board, such as parasitic resistances and capacitances. These parasitic components can often be estimated more accurately using SPICE simulation. Also, designers may want more information about the circuit than is available from a single mock-up. For instance, circuit performance is affected by component manufacturing tolerances. In these cases it is common to use SPICE to perform Monte Carlo simulations of the effect of component variations on performance, a task which is impractical using calculations by hand for a circuit of any appreciable complexity.

Circuit simulation programs, of which SPICE and derivatives are the most prominent, take a text netlist describing the circuit elements (transistors, resistors, capacitors, etc.) and their connections, and translate this description into equations to be solved. The general equations produced are nonlinear differential algebraic equations which are solved using implicit integration methods, Newton's method and sparse matrix techniques.

2. Transient Analysis

Since transient analysis is dependent on time, it uses different analysis algorithms, control options with different convergence-related issues and different initialization parameters than DC analysis. However, since a transient analysis first performs a DC operating point analysis (unless the UIC option is specified in the .TRAN statement), most of the DC analysis algorithms, control options, and initialization and convergence issues apply to transient analysis.

Some circuits, such as oscillators or circuits with feedback, do not have stable operating point solutions. For these circuits, either the feedback loop must be broken so that a DC operating point can be calculated or the initial conditions must be provided in the simulation input. The DC operating point analysis is bypassed if the UIC parameter is included in the .TRAN statement. If UIC is included in the .TRAN statement, a transient analysis is started using node voltages specified in an .IC statement. If a node is set to 5V in a .IC statement, the value at that node for the first time point (time 0) is 5V.

You can use the .OP statement to store an estimate of the DC operating point during a transient analysis.

.TRAN 1ns 100ns UIC .OP 20ns

The .TRAN statement UIC parameter in the above example bypasses the initial DC operating point analysis. The .OP statement calculates transient operating point at t = 20ns during the transient analysis.

Although a transient analysis might provide a convergent DC solution, the transient analysis itself can still fail to converge. In a transient analysis, the error message "internal timestep too small" indicates that the circuit failed to converge. The convergence failure might be due to stated initial conditions that are not close enough to the actual DC operating point values.

3. Program Features and Structure

SPICE became popular because it contained the analyses and models needed to design integrated circuits of the time, and was robust enough and fast enough to be practical to use. Precursors to SPICE often had a single purpose: The BIAS program, for example, did simulation of bipolar transistor circuit operating points; the SLIC (simulator for linear integrated circuits) program did only small-signal analyses. SPICE combined operating point solutions, transient analysis, and various small-signal analyses with the circuit elements and device models needed to successfully simulate many circuits.

3.1 Analyses

SPICE2 included these analyses:

(1) AC analysis (linear small-signal frequency domain analysis).

(2) DC analysis (nonlinear quiescent point calculation).

(3) DC transfer curve analysis (a sequence of nonlinear operating points calculated while sweeping an input voltage or current, or a circuit parameter).

(4) Noise analysis (a small signal analysis done using an adjoint matrix technique which sums uncorrelated noise currents at a chosen output point).

(5) Transfer function analysis (a small-signal input/output gain and impedance calculation).

(6) Transient analysis (time-domain large-signal solution of nonlinear differential algebraic equations).

Since SPICE is generally used to model nonlinear circuits, the small signal analyses are necessarily preceded by a quiescent point calculation at which the circuit is linearized. SPICE2 also contained code for other small-signal analyses: sensitivity analysis, pole-zero analysis, and small-signal distortion analysis. Analysis at various temperatures was done by automatically updating semiconductor model parameters for temperature, allowing the circuit to be simulated at temperature extremes.

Other circuit simulators have added many analyses beyond those in SPICE2 to address changing industry requirements. Parametric sweeps were added to analyze circuit performance with changing manufacturing tolerances or operating conditions. Loop gain and stability calculations were added for analog circuits. Harmonic balance or time-domain steady state analyses were added for RF and switched-capacitor circuit design. However, a public-domain circuit simulator containing the modern analyses and features needed to become a successor in popularity to SPICE has not yet emerged.

It is very important to use appropriate analyses with carefully chosen parameters. For example, application of linear analysis to nonlinear circuits should be justified separately. Also, application of transient analysis with default simulation parameters can lead to qualitatively wrong conclusions on circuit dynamics.

3.2   Device models

SPICE2 included many semiconductor device compact models: three levels of MOSFET model, a combined Ebers—Moll and Gummel-Poon bipolar model, a JFET model, and a model for a junction diode. In addition, it had many other elements: resistors, capacitors, inductors (including coupling), independent voltage and current sources, ideal transmission lines, active components and voltage and current controlled sources.

SPICE3 added more sophisticated MOSFET models, which were required due to advances in semiconductor technology. In particular, the BSIM family of models were added, which were also developed at UC Berkeley.

Commercial and industrial SPICE simulators have added many other device models as technology advanced and earlier models became inadequate. To attempt standardization of these models so that a set of model parameters may be used in different simulators, an industry working group was formed, the Compact Model Council, to choose, maintain and promote the use of standard models. The standard models today include BSIM3, BSIM4, BSIMSOI, PSP, HICUM, and MEXTRAM.

3.3   Input and output: Netlists, schematic capture and plotting

SPICE2 took a text netlist as input and produced line printer listings as output, which fit with the computing environment in 1975. These listings were either columns of numbers corresponding to calculated outputs (typically voltages or currents), or line-printer character "plots". SPICE3 retained the netlist for circuit description, but allowed analyses to be controlled from a command line interface similar to the C shell. SPICE3 also added basic X plotting, as UNIX and engineering workstations became common.

Vendors and various free software projects have added schematic capture front-ends to SPICE, allowing a schematic diagram of the circuit to be drawn and the netlist to be automatically generated. Also, graphical user interfaces were added for selecting the simulations to be done and manipulating the voltage and current output vectors. In addition, very capable graphing utilities have been added to see waveforms and graphs of parametric dependencies. Several free versions of these extended programs are available, some as introductory limited packages, and some without restrictions.

New Words

integrity	[ɪn'tegrəti]	n. 完整性
discrete	[dɪ'skri:t]	adj. 不连续的，分散的，离散的
breadboard	['bredbɔ:d]	n. 电路试验板
		vt. 为……制作；在平板上作……
		vi. 制作模拟板
photolithographic	[ˌfəutə,lɪθə'græfɪk]	adj. 照相平版印刷（法）的
parasitic	[ˌpærə'sɪtɪk]	adj. 寄生的
mock-up	['mɔk ʌp]	n. 实验或教学用的实物大模型
impractical	[ɪm'præktɪkl]	adj. 不切实际的，昧于实际的
complexity	[kəm'pleksəti]	n. 复杂性
derivative	[dɪ'rɪvətɪv]	adj. 引出的
		n. 派生的事物
netlist	['nɪtlɪst]	n. 连线表，网络表
convergence	[kən'vɜ:dʒəns]	n. 收敛
initialization	[ɪˌnɪʃəlaɪ'zeɪʃn]	n. 设定初值，初始化
statement	['steɪtmənt]	n. 语句
bypass	['baɪpɑ:s]	n. 旁路
		vt. 绕过，设旁路，迂回
convergent	[kən'vɜ:dʒənt]	adj. 会集于一点的，会聚性的，收敛的
robust	[rəʊ'bʌst]	adj. 健壮的
precursor	[prɪ'kɜ:sə]	n. 先驱
sweep	[swi:p]	v. 扫描
distortion	[dɪ'stɔ:ʃn]	n. 变形，失真
successor	[sək'sesə]	n. 继承者，接任者

default	[dɪˈfɔːlt]	n. 默认（值），缺省（值）
coupling	[ˈkʌplɪŋ]	n. 联结，接合，耦合
inadequate	[ɪnˈædɪkwət]	adj. 不充分的，不适当的
standardization	[ˌstændədaɪˈzeɪʃn]	n. 标准化
plot	[plɒt]	vt. 绘图
character	[ˈkærəktə]	n. 字符
workstation	[ˈwɜːksteɪʃn]	n. 工作站
graph	[ɡræf]	n. 图表，曲线图
linearize	[ˈlɪnɪəraɪz]	vt. 使线性化

Phrases

open source	开放源码，开源
analog electronic circuit	模拟电子电路
integrated circuit	集成电路
board-level design	板级设计
photolithographic mask	光刻掩模
printed wiring board	印制电路板
manufacturing tolerance	制造公差
Monte Carlo	蒙特卡罗
circuit element	电路元件
differential algebraic equation	微分代数方程
implicit integration methods	隐式积分方法
Newton's method	牛顿法
sparse matrix	稀疏矩阵
transient analysis	瞬态分析，暂态分析
initialization parameter	最初参数，初始参数
feedback loop	反馈回路，反馈环
of the time	当时的，当代的
bipolar transistor	双极晶体管
linear small-signal frequency domain analysis	线性小信号频域分析
nonlinear quiescent point calculation	非线性静态点计算
transfer curve	转换曲线
noise analysis	噪声分析
adjoint matrix	伴随矩阵
transfer function	传递函数，转移函数
time-domain large-signal solution	时域大信号解
sensitivity analysis	灵敏度分析
pole-zero analysis	极零点分析
temperature extremes	温度极限

parametric sweep	参数扫描
loop gain	环路增益
stability calculation	稳定计算，稳性计算
harmonic balance	谐波平衡
steady state	恒稳态，定态
compact model	精简模型，简化模型
junction diode	结型二极管
transmission line	传输线，波导线
semiconductor technology	半导体工艺
schematic capture	原理图捕获
line printer	行式打印机
command line	命令行
graphical user interface	图形用户界面

Abbreviations

SPICE (Simulation Program with Integrated Circuit Emphasis)	集成电路模拟程序
UIC (Use Initial Conditions)	使用初始条件
SLIC (Simulator for Linear Integrated Circuits)	线性集成电路模拟程序
JFER (Junction Field Effect Transistor)	结栅场效应晶体管
BSIM (Berkeley Short-channel IGFET Model)	伯克利短沟道IGFET模型

Exercises

【Ex.6】根据文章所提供的信息判断正误。

1. SPICE is a specific-purpose, open source analog electronic circuit simulator.
2. Board-level circuit designs can often be breadboarded for testing.
3. Board-level circuit designers need less information about the circuit than is available from a single mock-up.
4. You can use the .TRAN statement to store an estimate of the DC operating point during a transient analysis.
5. SPICE contained the analyses and models needed to design integrated circuits of the time.
6. SPICE is generally used to model linear circuits.
7. Loop gain and stability calculations were added for analog circuits.
8. Qualitatively wrong conclusions on circuit dynamics can result from application of transient analysis with default simulation parameters can lead to.
9. The standard models today include BSIM3, BSIM4, BSIMSOI, PSP, HICUM, and MEXTRAM.
10. SPICE2 produced line printer listings as output, and these listings were only columns of numbers corresponding to calculated outputs (typically voltages or currents).

科技英语翻译知识

从句的译法

英语的从句分为定语从句、主语从句、宾语从句、状语从句、表语从句及同位语从句。由于英汉两种语言结构的不同，因此翻译时，应该根据不同结构、不同含义采用不同译法。下面一一讨论。

1. 定语从句翻译法

1）合译法

把定语从句放在被修饰的词语之前，从而将英语复合句翻译成汉语单句。如：

(1) All substances which can conduct electricity are called conductors.

一切能导电的物质叫作导体。

（不能说："一切物质叫作导体，该物质能导电"。第一分句在意义上不能独立存在，所以要合起来翻译。）

(2) An electric motor changes electrical energy into mechanical energy that can be used to do work.

电动机能把电能变成可用来做功的机械能。

（此句的定语从句也可分开来译，但它较短，还是合译为好。）

(3) New electron tubes could be built that worked at much higher voltages.

如今制造的电子管，工作电压要高得多。

（这一句定语从句是全句的重点，将从句顺序译成简单句中的谓语，从而突出从句的内容。）

2）分译法

根据定语从句的不同情况，可以将其翻译成并列分句、其他从句或词组等。如：

(1) Mechanical energy is changed into electric energy, which in turn is changed into mechanical energy.

机械能转变为电能，而电能又转变为机械能。

（本句的翻译被拆成两个分句，变成转折分句。）

(2) This, of course, includes the movement of electrons, which are negatively charged particles.

当然，这包括电子（带负电的粒子）的运动。

（定语从句在本句的翻译中转变为词组。）

(3) The strike would prevent the docking of ocean steamships which require assistance of tugboats.

罢工会使远洋航船不能靠岸,因为它们需要拖船的帮助。
(翻译成原因状语从句。)

(4) A geological prospecting engineer who had made a spectral analysis of ores discovered a new open-cut coalmine.
一位地质勘探工程师对矿石进行了波谱分析之后,发现了新的露天煤矿。
(翻译为时间状语结构。)

(5) The delivery of public services has tended to be an area where we decorate an obsolete process with technology.
公共服务的提供方式已趋陈旧,这正是我们必须采用技术加以装备的领域。
(翻译为并列分句。)

(6) We now live in a very new economy, a service economy, where relationships are becoming more important than physical products.
现在我们正生活于一种全新的经济,即服务性经济中,各种关系比物质产品越来越重要。
(翻译为并列分句。)

2. 主语从句翻译法

1)"的"字结构

以 that、what、who、where、whatever 等代词引导的主语从句,可以将从句翻译成"的"字结构。如:

(1) It is important that science and technology be pushed forward as quickly as possible.
重要的是要把科学技术搞上去。

(2) Whoever breaks the law will be punished.
凡是犯法的人都要受到法律的制裁。
(主语从句与主句合译成简单句,按顺序译出。)

2)"主-谓-宾"结构

以上代词引导的主语从句也可以译成"主-谓-宾"结构,从句本身作句子的主语,其余部分按原文顺序译出。如:

Whether the Government should increase the financing of pure science at the expense of technology or vice versa often depends on the issue of which is seen as the driving force.
政府究竟是以牺牲对技术的经费投入来增加对纯理论科学的经费投入,还是相反,这往往取决于把哪一方看作是驱动的力量。

3)分译法

把原来的状语从句从整体结构中分离出来,译成另一个相对独立的单句。如:

It has been rightly stated that this situation is a threat to international security.
这个局势对国际安全是个威胁,这样的说法是完全正确的。
(It 是形式主语,that this situation is a threat to international security 是真正的主语。)

3．宾语从句翻译法

由 that、what、how、where 等词引导的宾语从句一般按照原文顺序翻译，即顺译法。如：

(1) Scientists have reason to think that a man can put up with far more radiation than 0.1 rem without being damaged.

科学家们有理由认为人可以忍受远超过 0.1 rem 的辐射而不受伤害。

(2) We wish to inform you that we specialize in the export of Chinese textiles and shall be glad to enter into business relations with you on the basis of equality and mutual benefit.

我公司专门办理中国纺织品出口业务，并愿在平等互利的基础上同贵公司建立业务关系。

4．状语从句翻译法

有时也可以译为并列句，如：

Electricity is such an important energy that modern industry could not develop without it.

电是一种非常重要的能量，没有它，现代化工业就不能发展。

（原文由 such…that…引导的结果状语从句译为汉语的并列句。）

5．表语从句翻译法

大部分情况下可以采用顺译法，也可以用逆译法。如：

(1) My point is that the frequent complaint of one generation about the one immediately following it is inevitable.

我的观点是一代人经常抱怨下一代人是不可避免的。

（顺译法）

(2) His view of the press was that the reporters were either for him or against him.

他对新闻界的看法是，记者们不是支持他，就是反对他。

（顺译法）

(3) Water and food is what the people in the area are badly needed.

该地区的人们最需要的是水和食品。

（逆译法）

6．同位语从句翻译法

先翻译从句，即从句前置。如：

(1) This is a universally accepted principle of international law that the territory sovereignty does not admit of infringement.

一个国家的领土不容侵犯，这是国际法中尽人皆知的准则。

(2) Despite the fact that comets are probably the most numerous astronomical bodies in the solar system aside from small meteor fragments and the asteroids, they are largely a mystery.

在太阳系中除小片流星和小行星外，彗星大概是数量最多的天体了，尽管如此，它们仍旧基本上是神秘莫测的。

Reading Material

阅读下列文章。

Text	Note

Electrical Engineering Software
 	
1. MATLAB (Software for Numerical Computing)
  MATLAB (MATrix LABarotary) is the most popular electrical engineering software among electrical engineering. It was launched in 1983 by Mathworks Inc. and was one of the first commercial packages for linear algebra. It has evolved over time and has become the most comprehensive[1] software for numerical computing, dynamic system simulations, algebraic solutions, symbolic[2] mathematics etc. It contains add-on packages (called toolboxes[3]) for various functionalities. Toolboxes provide built-in functions to perform numerical computations including but not limited to ordinary and partial differential[4] equations, optimization, linear system implementation, linear algebra, control system design, system identification, and curve fitting[5].
  The programming language used in MATLAB (the software package) is also called Matlab. Matlab is a high-level programming language, it contains a good number of built-in functions to efficiently deal with matrices[6], numerical computations, symbolic mathematics etc.
 	
2. Simulink (GUI-based Software for Dynamic System Simulation)
  Simulink is the GUI-based companion software for Matlab. It is powered by Matlab programming language. Many electrical engineers find Simulink much easier to use than MATLAB. When you use MATLAB and Simulink together, you combine textual and graphical programming to design your system in a simulation environment. You can directly use the thousands of algorithms that are already in MATLAB. You can use MATLAB to create input data sets to drive simulation and run thousands of simulations in parallel. You can also analyze and visualize[7] the data in MATLAB.
  Though Simulink is a general-purpose software for implementing graphical simulation, it has a specialized toolbox for simulating
	[1] *adj.* 综合的
	[2] *adj.* 符号的
	[3] *n.* 工具箱
	
	[4] *n.* 微分
	
	[5] curve fitting: 曲线拟合
	
	[6] *n.* 矩阵
	
	[7] *v.* 可视化，使可见

power systems. It can be used to simulate and analyze renewable energy resources, transmission lines, electrical transients, and standby[8] switching of power supply.

[8] *adj.* 备用的

3. Pspice (Electrical Schematic Software)

PSpice is an acronym for Personal Simulation Program with Integrated Circuit Emphasis. It is a modified version of the academically[9] developed SPICE, and was commercialized by MicroSim in 1984. MicroSim was purchased by OrCAD a decade later in 1998.

[9] *adv.* 学术地

OrCAD EE PSpice is a SPICE circuit simulatorapplication for simulation and verification of analog and mixed-signal circuits. It typically runs simulations for circuits defined in OrCAD Capture, and can optionally integrate with the MATLAB/Simulink, using the Simulink to PSpice Interface. OrCAD Capture and PSpice Designer together provide a complete circuit simulation and verification solution with schematic entry, native analog, mixed-signal, and analysis engines.

The PSpice advanced analysis simulation capabilities cover various analyses — sensitivity[10], Monte Carlo, smoke (stress), optimizer, and parametric plotter[11], and they provide in-depth understanding of circuit performance beyond basic validation.

[10] *n.* 敏感,（仪器等的）灵敏性

[11] *n.* 绘图机,绘图仪

4. Multisim (Circuit Simulation & PCB Design Software)

Multisim integrates industry-standard SPICE simulation with an interactive schematic environment to instantly visualize and analyze electronic circuit behavior. By adding powerful circuit simulation and analyses to the design flow, Multisim helps researchers and designers reduce printed circuit board (PCB[12]) prototype iterations and save development costs.

[12] 印制电路板

5. ETAP (An Electrical Engineering Software for Power Systems)

Being an industry-standard software, ETAP (Electrical Transient Analyzer Program) is a full spectrum[13] analytical electrical engineering software specializing in the analysis, simulation, monitoring, control, optimization, and automation of electrical power systems. The ETAP software offers the best and most comprehensive suite of integrated power system enterprise solution that spans from modeling to operation.

[13] *n.* 光谱,波谱;范围,系列

Various toolbars in ETAP provide almost all the analyses needed to design, regulation and operation of a power system. ETAP can be used to perform power flow analysis, relay coordination[14] and protection design, control system design, and optimal power flow.

[14] *n.* 协调

6. Power World Simulator (Visual Electrical Engineering Software software)

Power World Simulator is an interactive power system simulation package designed to simulate high voltage power system operation on a time frame ranging from several minutes to several days. The software contains a highly effective power flow analysis package capable of efficiently solving systems of up to 250,000 buses[15].

[15] *n.* 总线

The functionality of Power World Simulator can be increased by adding several additional add-on to the base simulator package. The add-ons can be used for distributed computing[16], adding the effect of Geomagnetically Induced Currents (GIC[17]), integrated topology processing, optimal power flow, transient stability, voltage stability and many more.

[16] distributed computing: 分布式计算

[17] 地磁感应电流

7. PSCAD (Electromagnetic Transient Analysis Software)

PSCAD is an electrical engineering software package for electromagnetic transient analysis in power systems. As power systems evolve, the need for accurate and intuitive simulation tools becomes more and more important. With PSCAD you can build, simulate, and model your systems with ease, providing limitless possibilities in power system simulation. It includes a comprehensive library of system models ranging from simple passive elements and control functions to electric machines and other complex devices.

8. PSS/E (An Electrical Engineering Software for Power System Simulations)

PSS/E stands for power system simulator / engineering. It is used by planning and operations engineers, consultants, universities, and research labs around the world. PSS/E is an excellent engineering oriented software for simulating and analyzing electromechanical[18] transient processes in power systems. PSSE allows you to perform a wide variety of analysis functions, including power flow, dynamics,

[18] *adj.* 电动机械的，机电的

short circuit, contingency analysis, optimal power flow, voltage stability, transient[19] stability simulation, and much more.

[19] *adj.* 瞬态的，短暂的

9. LabVIEW (Designing Interfacing and HMIs)

LabVIEW (Laborartory Virtual Instruments Engineering Workbench) is a systems engineering software for applications that require test, measurement, and control with rapid access to hardware and data insights.

The LabVIEW software offers a graphical programming approach that helps you visualize every aspect of your application, including hardware configuration[20], measurement data, and debugging. This visualization makes it simple to integrate measurement hardware from any vendor, represent complex logic on the diagram, develop data analysis algorithms, and design custom engineering user interfaces.

[20] *n.* 配置

For real-time control, LabVIEW is the best tool available in the market. It can connect with multiple devices to acquire data from sensors and control actuators based on processed data.

10. Keil μVision

For designing and testing embedded systems, microcontrollers are used extensively for control electrical instruments. Keil μVision provides an all-in-one[21] solution for programming embedded devices.

[21] *n.* 一体化，集成

The μVision IDE[22] combines project management, run-time environment, build tools, source code editing, and program debugging in a single powerful environment. μVision is easy to use and accelerates[23] your embedded software development. μVision supports multiple screens and allows you to create individual window layouts anywhere on the visual surface. The μVision Debugger provides a single environment in which you may test, verify, and optimize your application code. The debugger includes traditional features like simple and complex breakpoints[24], watch windows, and execution control and provides full visibility to peripheral[25] device.

[22] 集成开发环境

[23] *v.* 加快，加速

[24] *n.* 断点
[25] *adj.* 外围的

参 考 译 文

电子设计自动化

简而言之,电子设计自动化(EDA)是帮助设计人员创建和测试电子电路的过程。它可用于各种任务,包括原理图捕获、仿真、设计原型和布局印制电路板(PCB)。

1. 什么是 EDA?

EDA 是一类工程师用来设计和测试电子系统的软件工具。这些工具自动执行电子设计过程中的各种任务,如电路仿真、芯片的设计/布局和电路验证。

工程师使用 EDA 工具来创建集成电路(IC)、PCB 和现场可编程门阵列(FPGA)的设计。它们还用于创建单片系统(SoC)设计和嵌入式系统设计。简而言之,如果你看到了一个包含上述部件的电子设备,它很可能是使用电子设计自动化工具设计的。

最常见的 EDA 工具大致可分为以下三类:

(1)设计输入:用于创建描述设计的原理图或 HDL 代码。

(2)仿真:用于验证设计的功能。

(3)物理设计:用于将逻辑设计转换为物理布局。

EDA 工具用于整个设计过程,从创建电路的初始原理图,到模拟其行为,再到验证其是否满足所有必要的设计要求。近年来,由于电子设备的复杂性不断增加,因此 EDA 工具变得越来越强大和复杂。

设计人员现在严重依赖 EDA 工具来帮助他们验证设计的正确性并满足制造期限。如果没有 EDA 工具,几乎不可能及时且经济高效地创建当今复杂的电子设备。

2. EDA 如何工作?

从广义上讲,EDA 软件可帮助工程师设计、分析和优化电路元件和系统。通过自动化某些任务,如重复计算或元件的布局和布线,EDA 工具可以帮助工程师节省时间并提高准确性。市场上有许多不同类型的 EDA 工具,每种工具都旨在解决设计过程中特定阶段的问题。

三个主要阶段是:组件/系统级设计、电路级设计和布局级设计。

工程师通过组件/系统级设计来决定其电路或系统的整体架构。这包括定义每个组件的功能和接口及这些组件如何相互交互。

电路级设计是工程师对每个单独组件进行更详细的设计,包括指定电阻器、电容器、电感器等的值及设计连接这些组件的电路。

布局级设计是工程师获取前两个阶段的所有信息并将其转化为物理布局的过程。这通常是通过使用计算机辅助设计(CAD)软件创建电路板的二维或三维表示来完成的。

三个阶段完成后,最后一步是使用 EDA 仿真工具运行仿真来验证电路的行为是否符合预期。这些模拟允许工程师在各种条件下测试他们的设计，以确保它们满足所有规范。

在设计的每个阶段，EDA 自动执行各种任务，这些任务从简单的重复计算到更复杂的流程（如放置和布线组件），这为提高项目的准确性和及时性提供了宝贵的帮助。

3. 手动设计、CAD 和 EDA 的区别

手动设计正如它听起来的那样——产品设计师在没有任何软件或计算机的帮助下手动创建一切。直到 20 世纪末，这都是设计和构建电子产品的标准方法。虽然使用手动设计方法可以创造出高质量的产品，但这通常非常耗时且昂贵。

CAD 软件是计算机辅助设计软件，可帮助产品设计师创建其产品的二维模型和三维模型。它通常与手动设计方法结合使用，但也可以单独使用。一般来说，使用 CAD 软件设计的产品比完全手动设计的产品更容易生产且成本更低。

EDA 软件更进了一步，它把产品设计中的诸多任务自动化。除了帮助产品设计人员创建产品模型之外，EDA 软件还可以帮助完成管理组件库、模拟电路和生成制造文档等任务。由于 EDA 把产品设计中的诸多任务自动化了，因此它可以帮助企业节省大量时间和金钱。

4. 如何选择 EDA 工具？

EDA 工具可以帮助你节省时间、减少错误、优化性能并增强创造力。但是，有这么多可以选择的工具，如何为你的项目选择最佳的 EDA 工具呢？我们将探讨选择 EDA 工具时需要考虑的一些关键因素和功能。

4.1 项目范围和复杂性

选择 EDA 工具时首先要考虑的是项目的范围和复杂性。根据设计的类型、规模和详细程度，你可能需要不同的 EDA 工具来处理设计过程的不同方面。例如，如果你正在设计一个简单的逻辑电路，那么你可能只需要一个原理图编辑器和一个逻辑模拟器。但如果你正在设计复杂的 SoC，就可能需要一套 EDA 工具来处理高级综合、硬件描述语言、物理设计、验证和测试。

4.2 兼容性与集成度

选择 EDA 工具时要考虑的另一个重要因素是它们与其他工具和平台的兼容性和集成度。你希望确保 EDA 工具能够彼此良好地协同工作，并与可能在项目中使用的其他软件和硬件很好地协同工作。例如，如果你使用特定的微控制器或 FPGA，你需要确保你的 EDA 工具可以支持其编程和调试。或者，如果使用基于云的平台进行协作和存储，你希望确保你的 EDA 工具可以与其连接和同步。

4.3 可用性和功能

选择 EDA 工具时要考虑的第三个因素是它们的可用性和功能。你希望选择易于学习、使用和自定义的 EDA 工具，并提供项目所需的特性和功能。例如，如果你正在设计混合信号电路，你希望选择一个 EDA 工具，它能够同时处理模拟和数字的组件和信号。或者，如果你正在设计 PCB，则需要选择可以提供布局、布线和制造选项的 EDA 工具。

4.4 成本和可用性

选择 EDA 工具时要考虑的最后一个因素是其成本和可用性。你希望选择适合你的预算且易于使用的 EDA 工具。根据你的项目和偏好，可以选择免费或开源的 EDA 工具、商业或专有的 EDA 工具，以及两者的组合。你还需要检查 EDA 工具的许可和支持选项及其更新和升级。

5. EDA 的未来

EDA 领域一直在不断发展。以下是可能需要了解的 EDA 的一些重要趋势。

5.1 用于设计优化的机器学习

近年来，机器学习已应用于电子设计的很多不同领域，并取得了可喜的成果。机器学习可用于优化各种性能指标的设计，如功耗、速度和面积。由于设计师寻找事半功倍的方法，因此这种趋势在未来几年肯定会持续下去。

5.2 基于云的工具

云已经改变了许多行业的游戏规则，电子设计也不例外。与传统桌面工具相比，基于云的 EDA 工具有许多优势，如更低的成本、更轻松的协作和更好的可扩展性。随着越来越多的设计师涌向云，这种趋势只会持续下去。

5.3 集成模拟设计和数字设计

过去，模拟设计和数字设计是分开创建的，然后在制造前的最后一刻进行集成。然而，在当今快节奏的世界中，这种方法不再可行，因为上市时间至关重要。现在设计人员从一开始就将模拟设计和数字设计集成在一起，以创建更高效的工作流程并避免未来潜在的问题。

5.4 低功耗/节能设计方法

由于对能够在不插电的情况下长时间运行的移动设备的需求不断增长，功耗成为当今设计人员关注的主要问题。因此，低功耗/节能设计方法在当今的应用领域变得越来越流行。这些方法可帮助设计人员满足当今设备严格的功耗要求，同时仍保持消费者所需的性能水平。

5.5 使用自动化测试平台节省时间

为了保持竞争力，设计人员需要能够在不牺牲产品质量或性能水平的情况下尽快将其推向市场。实现这一目标的一种方法是在投入生产之前使用自动化测试平台来验证设计的功能。通过使用自动化测试平台，设计人员可以节省宝贵的时间，否则这些时间将用于手动测试其设计。他们可以将时间花在其他任务上，如优化或设计下一代产品。

EDA 领域在不断发展，这意味着今天被认为是前沿的东西到明天可能就会过时。然而，通过紧跟最新趋势，无论发生什么变化，都可以保持领先地位并让设计顺利进行。

Text A

Programmable Logic Controllers (PLCs)

1. What Is a PLC?

PLC stands for programmable logic controller. A PLC is a computer specially designed to operate reliably under harsh industrial environments—such as extreme temperatures, wet, dry, and/or dusty conditions. PLCs are used in almost every industry right from manufacturing, food processing, automotive, oil & gas, and many other industries.

2. How Does a PLC Work?

The working of a programmable logic controller can be easily understood as a cyclic scanning method known as the scan cycle.

A PLC scan process includes the following steps:

(1) The operating system starts cycling and monitoring the time.

(2) The CPU starts reading the data from the input module and checks the status of all the inputs.

(3) The CPU starts executing the user or application program written in relay-ladder logic or any other PLC-programming language.

(4) Next, the CPU performs all the internal diagnosis and communication tasks.

(5) According to the program results, it writes the data into the output module so that all outputs are updated.

This process continues as long as the PLC is in run mode.

3. Physical Structure of PLC

The structure of a PLC is almost similar to a computer's architecture. Programmable logic controllers continuously monitor the input values from various input sensing devices (e.g. accelerometer, weight scale, hardwired signals, etc.) and produce corresponding output depending on the nature of production and industry. A typical PLC consists of five parts, namely: rack or chassis, power supply module, central processing unit (CPU) module, input & output module and communication interface module.

3.1 Rack or chassis

In all PLC systems, the PLC rack or chassis forms the most important module and acts as a backbone to the system. PLCs are available in different shapes and sizes. When more complex control systems are involved, larger PLC racks are required.

3.2 Power supply module

This module is used to provide the required power to the whole PLC system. It converts the available AC power to the DC power which is required by the CPU and I/O module. [1] PLC generally works on a 24V DC supply.

3.3 CPU module

CPU module has a central processor, ROM & RAM. ROM is used to store an operating system, drivers, and application programs. RAM is used to store programs and data. CPU is the brain of PLC with an microprocessor. Being a microprocessor-based CPU, it replaces timers, relays, and counters. CPU reads the input data from sensors, processes it, and finally sends the command to controlling devices. CPU also contains other electrical parts to connect cables used by other units.

3.4 Input and output module

PLC has an exclusive module for interfacing inputs and output.

Input devices can be start and stop push buttons, switches, etc and output devices can be an electric heater, valves, relays, etc. I/O module helps to interface input and output devices with a microprocessor.

There are two main sections in the input module, namely the power section and the logical section. Both sections are electrically isolated from each other.

The output module of PLC works similarly to the input module but in the reverse process.

3.5 Communication interface module

To transfer information between CPU and communication networks, intelligent I/O modules are used. These communication modules help to connect with other PLCs and computers which are placed at a remote location.

4. Types of PLC

4.1 Mini PLCs

They are small, low-cost controllers that are ideal for simple control applications. They typically have fewer input/output (I/O) points than larger controllers and can be programmed using ladder logic or other programming languages. Mini PLCs offer fast installation due to their small size and often come with built-in I/O capabilities such as digital inputs, analog outputs, and pulse outputs.

4.2 Modular PLCs

They consist of a base unit that contains the processor module and communications ports, along with smaller modules that can be added to extend the system's functionality (see Figure 9-1). Modular systems offer more flexibility than fixed systems since they allow users to mix different types of I/O modules to meet their specific application requirements.

Figure 9-1　A Modular PLC

4.3 Fixed PLCs

They are designed for dedicated tasks and cannot easily be modified once installed. However, they provide cost-efficient solutions for many repetitive tasks. Fixed systems are suitable for straightforward process control applications where parameters do not need to change frequently or rapidly during operation.

4.4 Micro PLCs

They offer an intermediate level of complexity between mini models and modular models. They are usually compact devices capable of controlling multiple processes simultaneously without requiring additional hardware components like expansion cards or rack units found in some modular models (see Figure 9-2).[2] Microcontrollers can also feature integrated communication functions such as Ethernet networking protocols for easy integration into a distributed automation system architecture.

Figure 9-2　A Micro PLC

4.5 Nano PLCs

They represent the latest generation of programmable logic controllers. These ultra-compact devices use advanced microcontrollers combined with specialized programming software tools to reduce costs while providing fast processing speed and high accuracy. [3] Nano PLCs are even used in highly complex applications involving multiple axes movement or sophisticated machine vision operations like object recognition algorithms & pattern matching techniques.

4.6 Safety PLCs

A safety PLC is designed to implement safety functions in industrial automation (see Figure 9-3). It ensures the protection of personnel, equipment, and the environment by adhering to international safety standards. Safety PLCs incorporate redundancy and diagnostic capabilities, and are assigned safety integrity levels (SIL) to ensure high reliability and fault tolerance.

Figure 9-3　A Safety PLC

5. Types of PLC Output

5.1 Analog output

To control equipment that needs varying degrees of control, analog outputs are used to produce continuous voltage or current signals. Analog output offers precise control by altering the voltage or current levels they produce.

5.2 Relay output

The electrical separation between the PLC and the external devices is provided by relay outputs in PLCs. They are used to control machinery such as huge motors, contactors, or power relays that need larger current or voltage levels.

5.3 Transistor output

Transistor outputs are an output module in PLCs that use transistors as switching devices. Transistor outputs can handle higher currents and voltages compared to digital outputs, making them suitable for controlling devices that require more power.

6. PLC Programming

A PLC program consists of a set of instructions either in textual or graphical form, which represents the logic that governs the process the PLC is controlling. There are two main classifications of PLC programming languages, which are further divided into many sub-classified types[4].

(1) Textual language: instruction list, structured text.

(2) Graphical form: ladder diagrams (LD) (i.e. ladder logic), function block diagram (FBD), sequential function chart (SFC).

Although all of these PLC programming languages can be used to program a PLC, graphical languages (like ladder logic) are typically preferred to textual languages (like structured text programming).

6.1 Ladder logic

Ladder logic is the simplest form of PLC programming. It is also known as "relay logic". The relay contacts used in relay controlled systems are represented using ladder logic.

6.2 Functional block diagrams

Functional block diagram (FBD) is a simple and graphical method to program multiple functions in PLC. A function block is a program instruction unit that, when executed, yields one or more output values.

It is represented by a block as shown below (see Figure 9-4). It is represented as a rectangular block with inputs entering on left and output lines leaving at the right. It gives a relation between the state of input and output.

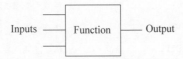

Figure 9-4 Function Block

The advantage of using FBD is that any number of inputs and outputs can be used on the functional block. When using multiple input and output, you can connect the output of one function block to the input of another whereby building a function block diagram (see Figure 9-5).

6.3 Structured text programming

Structured text is a textual programming language that utilizes statements to determine what to execute. It follows more conventional programming protocols but it is not case sensitive.

Structured text has a set of expressions, assignments, loops, conditional statements, and functions. Programmers can write reusable code blocks in a structured text language and use them in different programs. Some of the benefits of structured text are flexibility, reusability, readability, self-documenting and portability.

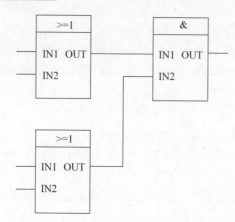

Figure 9-5　Example of Functional Block Diagram

New Words

programmable	[ˈprəʊɡræməbl]	adj.	可编程的
operate	[ˈɒpəreɪt]	v.	工作；运转；操作
harsh	[hɑːʃ]	adj.	严酷的；粗糙的；刺耳的
module	[ˈmɒdjuːl]	n.	模块；组件
diagnosis	[ˌdaɪəɡˈnəʊsɪs]	n.	诊断；判断
communication	[kəˌmjuːnɪˈkeɪʃn]	n.	通信
accelerometer	[əkˌseləˈrɒmɪtə]	n.	加速度计
rack	[ræk]	n.	支架
chassis	[ˈʃæsɪ]	n.	底架
interface	[ˈɪntəfeɪs]	n.	界面；接口
		v.	（使通过界面或接口）接合，连接
backbone	[ˈbækbəʊn]	n.	骨干，支柱
memory	[ˈmemərɪ]	n.	存储器，内存
microprocessor	[ˌmaɪkrəʊˈprəʊsesə]	n.	微处理器
replace	[rɪˈpleɪs]	v.	替换，取代；更新
timer	[ˈtaɪmə]	n.	定时器；计时器
relay	[ˈriːleɪ]	n.	继电器，中继设备
		v.	转播，转送
counter	[ˈkaʊntə]	n.	计数器
heater	[ˈhiːtə]	n.	加热器
isolate	[ˈaɪsəleɪt]	vt.	使隔离；使绝缘
		vi.	隔离，孤立
remote	[rɪˈməʊt]	adj.	远程的
		n.	遥控（器）
low-cost	[ˈləʊ ˈkɒst]	adj.	价格便宜的，廉价的
installation	[ˌɪnstəˈleɪʃn]	n.	安装；装置

capability	[ˌkeɪpəˈbɪlətɪ]	n. 能力，性能
functionality	[ˌfʌŋkʃəˈnælətɪ]	n. 功能；功能性
cost-efficient	[ˌkɒst ɪˈfɪʃənt]	adj. 有成本效益的，划算的
repetitive	[rɪˈpetətɪv]	adj. 重复的
straightforward	[ˌstreɪtˈfɔːwəd]	adj. 简单明了的
parameter	[pəˈræmɪtə]	n. 参数，参量
intermediate	[ˌɪntəˈmiːdɪət]	adj. 中间的，中级的 n. 中间物，中间人
compact	[kəmˈpækt]	adj. 紧凑的；简洁的
Ethernet	[ˈiːθənet]	n. 以太网
protocol	[ˈprəʊtəkɒl]	n. 协议
distribute	[dɪˈstrɪbjuːt]	vt. 分布，分配，散布，分发
nano	[ˈnænəʊ]	n. 纳米技术，毫微技术
recognition	[ˌrekəgˈnɪʃn]	n. 识别，认识
algorithm	[ˈælgərɪðəm]	n. 算法
protection	[prəˈtekʃn]	n. 保护
redundancy	[rɪˈdʌndənsɪ]	n. 冗余
precise	[prɪˈsaɪs]	adj. 精确的；清晰的
handle	[ˈhændl]	v. 处理；操作，操控
textual	[ˈtekstʃuəl]	adj. 文本的
classification	[ˌklæsɪfɪˈkeɪʃn]	n. 分类；分级；类别
rectangular	[rekˈtæŋɡjələ]	adj. 矩形的
expression	[ɪkˈspreʃn]	n. 表示式
assignment	[əˈsaɪnmənt]	n. 赋值
loop	[luːp]	n. 循环
programmer	[ˈprəʊɡræmə]	n. 程序员，程序设计者
reusable	[ˌriːˈjuːzəbl]	adj. 可再用的，可重用的
reusability	[riːjuːzəˈbɪlɪtɪ]	n. 可重用性

Phrases

industrial environment	工业环境
cyclic scanning method	循环扫描法
operating system	操作系统
input module	输入模块
application program	应用程序
relay-ladder logic	继电器梯形逻辑
according to	根据……
output module	输出模块
as long as	只要……

sensing device	灵敏元件，传感器
power supply module	电源模块，供电模块
control system	控制系统
convert ... to ...	把……转换为……
communication network	通信网络
programming language	编程语言
pulse output	脉冲输出
communications port	通信端口
expansion card	扩充插件板，扩展卡
ultra-compact device	超紧凑装置
machine vision	机器视觉
object recognition	目标识别，物体识别
pattern matching	模式匹配
fault tolerance	容错（性）
electrical separation	电子隔离
external device	外部设备
textual language	文本语言
graphical form	图形形式
ladder logic	梯形逻辑
structured text programming	结构化文本编程
relay logic	继电器逻辑
case sensitive	区分大小写
conditional statement	条件语句

Abbreviations

PLC (Programmable Logic Controller)	可编程逻辑控制器
CPU (Central Processing Unit)	中央处理器
I/O (Input/Output)	输入/输出
ROM (Read Only Memory)	只读存储器
RAM (Random Access Memory)	随机存储器
SIL (Safety Integrity Level)	安全完整性等级
LD (Ladder Diagram)	梯形图
FBD (Function Block Diagram)	功能块图
SFC (Sequential Function Chart)	顺序功能图

Notes

[1] It converts the available AC power to the DC power which is required by the CPU and I/O module.

本句中，which is required by the CPU and I/O module 是一个定语从句，修饰和限定 the

DC power。

本句意为：它将可用的交流电转换为 CPU 和输入/输出模块所需的直流电。

[2] They are usually compact devices capable of controlling multiple processes simultaneously without requiring additional hardware components like expansion cards or rack units found in some modular models.

本句中，capable of controlling multiple processes simultaneously 是一个形容词短语，作后置定语，修饰和限定 compact devices。它可以扩展为一个定语从句：which are capable of controlling multiple processes simultaneously。be capable of 的意思是"能够"。found in some modular models 是一个过去分词短语，作定语，修饰和限定 expansion cards or rack units。它也可以扩展为一个定语从句：which are found in some modular models。

本句意为：它们通常是紧凑的设备，能够同时控制多个进程，而不需要额外的硬件组件（如某些模块化模型中的扩展卡或机架单元）。

[3] These ultra-compact devices use advanced microcontrollers combined with specialized programming software tools to reduce costs while providing high levels of processing speed and accuracy.

本句中，combined with specialized programming software tools 是一个过去分词短语，作定语，修饰和限定 advanced microcontrollers。combined with 表示"合并在一起、结合在一起"。to reduce costs 是一个动词不定式短语，作目的状语。while providing high levels of processing speed and accuracy 作时间状语。while 表示两件事同时发生，意思是"与……同时"。

本句意为：这些超紧凑的设备使用了先进的微控制器，与专门的编程软件工具相结合，以降低成本，同时保持处理速度快且准确性高。

[4] There are two main classifications of PLC programming languages, which are further divided into many sub-classified types.

本句中，which are further divided into many sub-classified types 是一个非限定性定语从句，对 two main classifications of PLC programming languages 进行补充说明。

本句意为：PLC 编程语言主要有两大类，并进一步分为许多子分类类型。

Exercises

【Ex.1】根据课文内容，回答以下问题。

1. What does PLC stand for? What is a PLC?

2. How many parts does a typical PLC consist of ? What are they?

3. How many types of PLCs are mentioned in the passage? What are they?

4. How many types of PLC output are mentioned in the passage? What are they?

5. What does a PLC program consists of?

【Ex.2】把下面的单词填写到合适的空格中。

| A | magnitude | B | terminals | C | elements | D | charge | E | increase |
| F | measuring | G | conductors | H | fuse | I | Reclosers | J | interrupts |

Electrical Substation

An electrical substation consists of the following parts.

Transformer: it is a static electrical machine that serves to ___1___ or decrease electricity in an AC electrical circuit, while maintaining a constant frequency and power.

Circuit breaker: it ___2___ and reestablishes the continuity of an electric circuit. Such interruption is made with load or short-circuit current.

Recloser: it is an electromechanical part that interrupts the current when there is an excess of electricity and acts when a fault is generated in the circuit. ___3___ are designed to operate with 3 closing operations and 4 openings with an interval between them.

Blade fuses: they are connection and disconnection ___4___ of electric circuits with a double function. On the one hand, as a blade disconnect, it switches on and off. On the other hand, it acts as a ___5___ protection element and is used when an overcurrent is registered.

Disconnect switches and test switches: they serve to physically disconnect an electric circuit, so they usually operate without ___6___. These switches work mechanically and also manually.

Lightning arresters: they are responsible for keeping ionized rays away. When there is a surge of a certain ___7___, lightning arresters form an electronic arc that makes the current discharge on the ground and not on people or equipment and installations.

Instrument transformers: they are apparatuses responsible for ___8___ the electric current.

Junction boxes: they are the connection ___9___ per phase that allow us to make derivations and to reach specific areas.

Condensers: they allow us to conserve the electricity that is produced in an electric field. Through two ___10___ separated by insulating material, energy is temporarily stored.

【Ex.3】把下列句子翻译成中文。

1. Programmable logic controller (PLC) is a control system using electronic operations.

2. In the event of the machine not operating correctly, an error code will appear.

3. The system includes core module and application module.

4. This information is no longer retained within the computer's main memory.

5. Companies like Intel developed faster microprocessors, so personal computers could process the incoming signals at a more rapid rate.

6. Relays are highly versatile components that are just as effective in complex circuits as in simple ones.

7. You must configure the operating system to proceed with the installation.

8. This laptop is significantly more compact than any comparable laptop, with no loss in functionality.

9. There are some goals of certified E-mail protocol: confidentiality, non-repudiation and fairness.

10. Experimental data shows the authentication algorithm is correct and effective.

【Ex.4】把下列短文翻译成中文。

The function block diagram (FBD) is a graphical language for programmable logic controller design. It can describe the function between input variables and output variables. A function is described as a set of elementary blocks. Input and output variables are connected to blocks by connection lines.

Inputs and outputs of the blocks are wired together with connection lines or links. Single lines may be used to connect two logical points of the diagram:

(1) An input variable and an input of a block.

(2) An output of a block and an input of another block.

(3) An output of a block and an output variable.

The connection is oriented, meaning that the line carries associated data from the left end to the right end. The left and right ends of the connection line must be of the same type.

Multiple right connection can be used to broadcast information from its left end to each of its right ends. All ends of the connection must be of the same type.

【Ex.5】通过 Internet 查找资料，借助电子词典、辅助翻译软件及 AI 工具，完成以下技术报告，并附上收集资料的网址。通过 E-mail 发送给老师，或按照教学要求在网上课堂提交。
1. 当前世界上有哪些最主要的 PLC 生产厂家及有哪些最新型的产品（附各种最新产品的图片）。
2. 当前有哪些最主要的 PLC 编程语言（附各种编程语言的操作界面）。

Text B

Smart Power Grids

1. What Is a Smart Power Grid?

A smart power grid simply refer to an electrical grid powered with modern technologies, such as big data, sensors, wireless modules, and monitoring systems to make the grid more efficient and resilient.

Smart power grids can collect real-time information on electricity consumption, demand, and any issues spotted in the grid and then send the information to suppliers and consumers. Through the sensors, the power grids can also monitor the activities of all devices connected to the grind and even give suggestions to consumers on how they can use their energy more efficiently. Not only can they make consumers aware of how much energy they are using, but also alert them to any issue or malfunction that could cause a power outage.

But smart power grids also come with a few more useful features:

(1) Self-healing, isolating all malfunctioning parts of the grids so they wouldn't damage other parts.

(2) Demand response, through which consumers are recommended to take actions to reduce the load on the grid during peak hours and this way, stabilizing the demand.

(3) Support for emerging technologies, such as renewable energy sources or electric vehicles.

2. Why Do We Need Smarter Power Grids?

The conventional power grid has been in use since 1896 and has truly revolutionized our world. However, after decades of working well, the traditional infrastructure is now dealing with several issues:

(1) The aging infrastructure can barely handle the growing demand for electricity.

(2) Connecting renewable energy sources such as wind turbines or solar panels to the traditional grid is complicated.

(3) The infrastructure is getting more and more vulnerable to severe weather conditions or power overload, which means frequent outages and blackouts.

(4) Consumers don't have enough information on how they could cut their power consumption and lower their bills, as the billing can only give them an estimation of the energy they used during the month.

(5) The costs of protecting and repairing the infrastructure are increasing.

Smart power grid technologies can not only help us to solve those problems but also pave the way for a cleaner, more efficient, and affordable energy supply.

3. What Are the Benefits of Smart Power Grids?

3.1 Improving energy efficiency

Smart power grids can help enhance our energy efficiency in a few ways, thanks to automatically gathering and acting on data to manage demand and reduce wastage. For example, the grids can analyze what the peak usage hours are and then schedule power-demanding activities to run when the energy demand is the lowest.

This small change allows consumers to save a significant amount of money since they can now immediately find out how much they will pay for charging their EV car or running a dishwasher during the day and during the night.

3.2 Less time needed to restore power

Smart power grid technology can also help energy suppliers find out which areas are down if there is an outage. The in-built sensors will keep monitoring the entire infrastructure for any issues, and when they notice a problem, they will mark the area and then pass the information to the electricians. The team then gets plenty of information on where is the source of the issue and what might have caused it, making it easier for them to fix the problem.

The grids can also isolate the affected area to prevent the outage from spreading and reduce the number of consumers that will be affected by the blackout. This feature can drastically lower the number of wide-area blackouts happening, especially as those are both incredibly costly and also require a lot of time to be fixed.

3.3 A great way to fight electricity theft

Together with the growing electricity prices, the number of people attempting to steal electricity is, unfortunately, increasing as well. The losses mainly come from consumers who are tampering with the energy meters to slow them down or stop them altogether. And as it is impossible to monitor every single conventional meter, there is little the power providers can do to fight theft cases.

Tampering with smart power grids can prove far riskier for consumers though. As modern grids can keep collecting, updating, and sharing data on consumption remotely, companies can monitor the grid's activity without needing to track every separate meter. In addition, in case there are any unusual changes in consumer energy usage (such as a sudden drop in how much they are consuming), the meters will immediately highlight it to the suppliers, allowing them to investigate the situation straight away.

3.4　Promoting renewable energy usage

Modern grids can also encourage turning to renewable energy, simply by making it much easier for consumers to add and use those for powering their homes or workplace.

For example, smart power grids make it possible for consumers to track how much energy they are generating from solar panels and decide how they want to use it. Another incredible benefit of those is that they can connect many separate power plants (wind turbines, solar panels, fossil plants, etc.) and automatically switch to a different power source in case one of them is down.

As they can also manage energy storage and release surplus power when needed, smart power grids can ensure a stable power supply even when wind turbines or solar panels generate less power.

3.5　Reduced maintenance and repair costs

Maintenance and repair of the aging infrastructure also take far more time and money than they used to. First of all, the current infrastructure is pretty vulnerable to severe weather conditions, which has become more frequent in recent years. But there's also the problem that our energy needs grow beyond conventional grids' capabilities, leading to frequent grid overloads and outages. Due to the way the infrastructure is connected though, one power outage can cause a domino effect that can cause large-scale blackouts. The larger the area affected by the blackouts, the costlier it is to repair the infrastructure.

Smart power grids have a unique feature to lower the risk (and costs) of outages through self-healing. The self-healing feature allows the grids to immediately detect any malfunctioning parts of the network and then to isolate this part from the rest of the system to prevent other parts from being damaged. In addition, 24/7 infrastructure monitoring, thanks to multiple sensors, will also allow consumers to spot any problems with the infrastructure (such as malfunctioning devices), making it easier to prevent power outages and power cuts.

3.6　More accurate billing

Finally, smart power grids will allow consumers to know exactly how much energy they used during the month and only to pay for what they used. As they can check their usage at any time (down to how much energy each of their appliances is using), it will become much easier for them to find out where they might be wasting energy and reduce their consumption.

4. Do Smart Power Grids Have Any Disadvantages?

There are also a few issues that still prevent modern grids from becoming widespread:

Smart power grids are still fairly expensive: the smart power grids is made of multiple different elements that need to work together smoothly for the grids to work to their full potential. That means that the installation will take some time and, unfortunately, the upfront costs can be quite high as well.

Data privacy and security concerns: people are also wary of the intelligent grids because they see the collecting and sharing of their energy usage data as a violation of their privacy. Many of them also worry that the infrastructure could be hacked, and then their sensitive information will be leaked outside.

The technology isn't regulated yet: as the smart power grid technology is relatively new, there are not enough regulatory norms for the technology yet, whether in terms of installation, security, or privacy compliance.

New Words

resilient	[rɪˈzɪlɪənt]	adj. 有弹性的，复原的
collect	[kəˈlekt]	vt. 收集
spot	[spɒt]	v. 发现
suggestion	[səˈdʒestʃən]	n. 建议，意见
alert	[əˈlɜːt]	adj. 警觉的，警惕的，注意的
		n. 警报
		vt. 向……报警
malfunction	[ˌmælˈfʌŋkʃn]	vi. 失灵；发生故障
		n. 故障；功能障碍；失灵
self-healing	[self ˈhiːlɪŋ]	adj. 自我修复的
stabilize	[ˈsteɪbəlaɪz]	vt. （使）稳定，（使）稳固
infrastructure	[ˈɪnfrəstrʌktʃə]	n. 基础设施，基础建设
aging	[ˈeɪdʒɪŋ]	n. 老化
turbine	[ˈtɜːbaɪn]	n. 涡轮机，汽轮机，透平机
vulnerable	[ˈvʌlnərəbl]	adj. 易受攻击的
overload	[ˌəʊvəˈləʊd]	vt. 使超载；超过负荷
		n. 过量，超负荷
blackout	[ˈblækaʊt]	n. 停电
protect	[prəˈtekt]	vt. 保护
pave	[peɪv]	vt. 铺设，为……铺平道路
wastage	[ˈweɪstɪdʒ]	n. 消耗
restore	[rɪˈstɔː]	vt. 恢复，修复

electrician	[ɪˌlekˈtrɪʃn]	n. 电工，电气技师
drastically	[ˈdrɑːstɪklɪ]	adv. 大大地，彻底地
investigate	[ɪnˈvestɪgeɪt]	vt. 调查，审查
situation	[ˌsɪtʃuˈeɪʃn]	n. 情况
encourage	[ɪnˈkʌrɪdʒ]	vt. 鼓励，支持，促进
workplace	[ˈwɜːkpleɪs]	n. 工作场所，车间；工厂
incredible	[ɪnˈkredəbl]	adj. 不可思议的；惊人的；难以置信的
surplus	[ˈsɜːpləs]	adj. 过剩的；多余的 n. 剩余额
widespread	[ˈwaɪdspred]	adj. 分布广的；普遍的
hack	[hæk]	v. 非法侵入，攻破
regulatory	[ˈregjələtərɪ]	adj. 监管的；调整的
norm	[nɔːm]	n. 标准；规范；准则
compliance	[kəmˈplaɪəns]	n. 合规性

Phrases

smart power grid	智能电网
big data	大数据
monitoring system	监控系统
real-time information	实时信息
power outage	停电，断电
peak hours	高峰时间
emerging technology	新兴技术，新兴科技
renewable energy source	可再生能源
electric vehicle	电动车辆
solar panel	太阳电池板
plenty of	很多，大量
energy meter	电表
repair cost	修理成本，修缮费用
weather condition	气候条件，天气条件
domino effect	多米诺效应
large-scale blackout	大规模停电
power cut	停电，切断电源
upfront cost	前期成本，预付成本
a violation of	违反
sensitive information	敏感信息

Exercises

【Ex.6】根据文章所提供的信息判断正误。

1. A smart power grid simply refer to an electrical grid powered with modern technologies.
2. Through the sensors, the power grids can monitor the activities of all devices connected to the grind, but they can not give suggestions to consumers on how they can use their energy more efficiently.
3. The conventional power grid has been in use since 1869 and has truly revolutionized our world.
4. The costs of protecting and repairing the conventional infrastructure are increasing.
5. Smart power grids can analyze what the peak usage hours are and then schedule power-demanding activities to run when the energy demand is the lowest.
6. Smart power grid technology can not help energy suppliers find out which areas are down if there is an outage.
7. Power companies can monitor the grid's activity without needing to track every separate meter because modern grids can keep collecting, updating, and sharing data on consumption remotely.
8. The self-healing feature allows the grids to immediately detect any malfunctioning parts of the network, but they can not isolate this part from the rest of the system.
9. Smart power grids are still fairly expensive.
10. Though the smart power grid technology is relatively new, there are enough regulatory norms for the technology.

科技英语翻译知识

汉语四字格的运用

汉语的四字格，既包括结构严密、不能随意拆开的四言成语，也包括任意组合而成的四字词组。这种以四字格为基本形式的四字词组，言简意赅，形象生动，音节优美，韵律协调。再加上四字格结构灵活多变，它几乎能配置任何一种语法关系，满足结构变化的需要，所以汉语成语有97%采用四字格。在不影响意义表达的情况下，普通的五字词语和三字词语也经常用四字格形式代替。英译汉时，四字格若使用恰当，则既能保存原作的风姿，又能使译文大为生色，在进行科技英语翻译时也应该在忠实原文的基础上提倡使用四字格。例如：

(1) Owing to the frequency of collisions between molecules, their motions are entirely at random.

译文一　由于分子之间碰撞频繁，分子运动完全是杂乱的。

译文二　由于分子之间碰撞频繁，分子运动完全是杂乱无章的。

(2) The new computers are indeed cheap and fine.

译文一　这些新计算机确实又便宜又好。

译文二　这些新计算机确实物美价廉。

(3) Copper is an important conductor, both because of its high conductivity and because of its abundance and low cost.

译文一　铜是一种重要导体，因为它的电导率高，而且数量多，价格低。

译文二　铜是一种重要导体，因为它的电导率高，而且资源丰富，价格低廉。

(4) When we speak, sound waves begin to travel and go in all directions.

译文一　我们说话时，声波便开始向周围扩散。

译文二　我们说话时，声波便开始向四面八方扩散。

(5) In the long run, basic knowledge and technological application go hand in hand—one helps the other.

译文一　从长远来看，基础知识和技术应用是相互一致的。

译文二　从长远来看，基础知识和技术应用是携手并进、相辅相成的。

(6) A large segment of mankind turns to untrammeled nature as a last refuge from encroaching technology.

译文一　许多人都想寻找一块自由的地方，作为他们躲避现代技术侵害的避难所。

译文二　许多人都想寻找一块自由自在的地方，作为他们躲避现代技术侵害的世外桃源。

(7) It is evident that oxygen is the most active element in the atmosphere.

译文一　很显然，氧是大气中最活泼的元素。

译文二　毋庸置疑，氧是大气中最活泼的元素。

(8) All bodies on the earth are known to possess weight.

译文一　大家都知道，地球上的一切物质都有重量。

译文二　众所周知，地球上的一切物质都有重量。

(9) Wireless technology usually suffers when compared to plain-old cable.

译文一　无线电技术和普通电缆相比往往就不利了。

译文二　无线电技术和普通电缆相比往往就相形见绌了。

(10) Electrons and protons carry equal but opposite electrical charges, the number of electrons in the shell of a normal atom is equal to the number of protons in the nucleus.

译文一　电子和质子带有相等但相反的电荷，普通原子外围的电子数与原子核中的质子数相同。

译文二　电子和质子携带数目相等、符号相反的电荷，普通原子外围的电子数正好等于原子核中的质子数。

从以上 10 个例句可以看出，每个例句的第一个译文虽然都将意思表达正确，但是都是平铺直叙，形式上不工整，音节上更谈不上优美，读来没有回味。而第二个译文，在不改变原意的基础上，由于准确使用四字格，言简意赅，节奏感强，形式工整，既表达了原意，又增加了美感，读起来具有感染力。

但是在提倡运用四字格的同时，也要避免乱用四字格，否则会适得其反，以词害义。例如：

This aircraft is small, cheap, pilotless.

译文一　这种飞机小巧玲珑，价廉物美，无人驾驶。
译文二　这种飞机体积不大，价格便宜，无人驾驶。

原文体积上说到小，但没有说"玲珑"，价格上说到廉，但没有说"物美"。译文一形式上均匀对称，但是表达的意思超出了原文的范围，因而是不正确的翻译。这时就要宁可舍弃形式上的优美，而要达到词义的准确，如译文二。

总之，四字格的运用是翻译中一个重要的修辞方法。因为四字格可以使译文更加传神达意，增加译文的风采。但是，重要的一点是，四字格运用是以保证原意的传达为基础的，切不可一味追求语言上的美感，而不顾原文的含义乱用四字格，否则会产生相反的效果。

Reading Material

阅读下列文章。

Text	Note
VHDL **1. What Is VHDL** 　　VHDL is the VHSIC[1] Hardware Description Language. VHSIC is an abbreviation for very-high speed integrated circuit. It can describe the behaviour and structure of electronic systems, but is particularly suited as a language to describe the structure and behaviour of digital electronic hardware designs, such as ASICs[2] and FPGAs[3] as well as conventional digital circuits. 　　VHDL is a notation[4], and is precisely and completely defined by the language reference manual (LRM). This sets VHDL apart from other hardware description languages, which are to some extent defined in an ad hoc way by the behaviour of tools that use them. VHDL is an international standard, regulated by the IEEE[5]. The definition of the language is non-proprietary. 　　VHDL is not an information model, a database schema, a simulator, a toolset or a methodology[6]! However, a methodology and a toolset are essential for the effective use of VHDL. 　　Simulation and synthesis[7] are the two main kinds of tools which operate on the VHDL language. The Language Reference Manual does not define a simulator, but unambiguously[8] defines what each simulator must do with each part of the language. 　　VHDL does not constrain the user to one style of description. VHDL allows designs to be described using any methodology—top down, bottom up or middle out! VHDL can be used to describe	[1] 超高速集成电路 [2] 专用集成电路 [3] 现场可编程门阵列 [4] *n*. 符号 [5]（美国）电气电子工程师协会 [6] *n*. 方法学，方法论 [7] *n*. 综合，合成 [8] *adv.* 明白地，不含糊地

hardware at the gate level or in a more abstract[9] way. Successful high level design requires a language, a tool set and a suitable methodology. VHDL is the language, you choose the tools, and the methodology.

2. A Brief History of VHDL

2.1 The requirement

The development of VHDL was initiated[10] in 1981 by the United States Department of Defence to address the hardware life cycle crisis. As technologies became obsolete[11] the cost of reprocuring electronic hardware was reaching crisis point because the function of the parts was not adequately documented, and the various components making up a system were individually verified using a wide range of different and incompatible simulation languages and tools. The requirement for a language was that the language should have a wide range of descriptive capability that would work the same on any simulator and was independent of technology or design methodology.

2.2 Standardization

The standardization process for VHDL was unique in that the participation[12] and feedback from industry were sought at an early stage. A baseline language (version 7.2) was published before the standard so that tool development could begin in earnest in advance of the standard.

2.3 VHDL review

As an IEEE standard, VHDL must undergo a review process every 5 years (or sooner) to ensure its ongoing relevance to the industry.

3. Levels of Abstraction

VHDL can be used to describe electronic hardware at many different levels of abstraction. When considering the application of VHDL to FPGA/ASIC design, it is helpful to identify and understand the three levels of abstraction shown opposite—algorithm, register transfer level (RTL[13]), and gate level. Algorithms are unsynthesizable,

[9] *adj.* 抽象的，深奥的，理论的

[10] *vt.＆vi.* 开始，发起
[11] *adj.* 陈旧的

[12] *n.* 分享，参与

[13] 电阻-晶体管逻辑

RTL is the input to synthesis, gate level is the output from synthesis. The difference between these levels of abstraction can be understood in terms of timing.

3.1 Algorithm

A pure algorithm consists of a set of instructions that are executed in sequence to perform some task. A pure algorithm has neither a clock nor detailed delays. Some aspects of timing can be inferred from the partial ordering of operations within the algorithm. Some synthesis tools (behavioural[14] synthesis) are available that can take algorithmic VHDL code as input. However, even in the case of such tools, the VHDL input may have to be constrained[15] in some artificial[16] way, perhaps through the presence of an "algorithm" clock—operations in the VHDL code can then be synchronized to this clock.

[14] *adj.* 行为的，动作的
[15] *adj.* 被强迫的
[16] *adj.* 人造的

3.2 RTL

An RTL description has an explicit clock. All operations are scheduled to occur in specific clock cycles, but there are no detailed delays below the cycle level. Commercially available synthesis tools do allow some freedom in this respect. A single global clock is not required but may be preferred. In addition, retiming is a feature that allows operations to be re-scheduled across clock cycles, though not to the degree permitted in behavioural synthesis tools.

3.3 Gates

A gate level description consists of a network of gates and registers instanced from a technology library, which contains technology—specific delay information for each gate.

3.4 Writing VHDL for Synthesis

In the diagram above, the RTL level of abstraction is highlighted. This is the ideal level of abstraction at which to design hardware given the state of the art of today's synthesis tools. The gate level is too low a level for describing hardware—remember we're trying to move away from the implementation concerns of hardware design, we want to abstract to the specification level—what the hardware does, not how it does it. Conversely[17], the algorithmic level is too high a level, most commercially available synthesis tools cannot produce

[17] *adv.* 倒向地，逆向地，相反地

hardware from a description at this level.

In the future, as synthesis technology progresses, we will one day view the RTL level of abstraction as the "dirty" way of writing VHDL for hardware and writing algorithmic (often called behavioural) VHDL will be the norm[18].

Until then, VHDL coding at RTL for input to a synthesis tool will give the best results. Getting the best results from your synthesizable RTL VHDL is a key topic of the Comprehensive VHDL and Advanced VHDL Techniques training courses. The latter also covers behavioural synthesis techniques.

4. Scope of VHDL

VHDL is suited to the specification, design and description of digital electronic hardware.

4.1 System level

VHDL is not ideally suited for abstract system-level simulation, prior to the hardware-software split. Simulation at this level is usually stochastic[19], and is concerned with modelling performance, throughput, queueing[20] and statistical distributions. VHDL has been used in this area with some success, but is best suited to functional and not stochastic simulation.

4.2 Digital

VHDL is suitable for use today in the digital hardware design process, from specification through high-level functional simulation, manual design and logic synthesis down to gate-level simulation. VHDL tools usually provide an integrated design environment in this area.

VHDL is not suited for specialized implementation-level design verification tools such as analog simulation, switch level simulation and worst case timing simulation. VHDL can be used to simulate gate level fan out loading effects providing coding styles are adhered to and delay calculation tools are available. The standardization effort named VITAL (VHDL Initiative Toward ASIC Libraries) is active in this area, and is now bearing fruit in that simulation vendors have built-in VITAL support. More importantly, many ASIC vendors have VITAL-compliant libraries, though not all are allowing VITAL-based sign-off,not yet anyway.

[18] *n.* 标准，规范

[19] *adj.* 随机的
[20] *n.* 排队

4.3 Analogue

Because of VHDL's flexibility as a programming language, it has been stretched[21] to handle analog and switch level simulation in limited cases. However, look out for future standards in the area of analog VHDL.

4.4 Design process

The diagram below shows a very simplified view of the electronic system design process incorporating VHDL. The central portion of the diagram shows the parts of the design process which are most impacted[22] by VHDL.

5. Design Flow using VHDL

The diagram below summarizes the high level design flow for an ASIC (i.e. gate array[23], standard cell) or FPGA. In a practical design situation, each step described in the following sections may be split into several smaller steps, and parts of the design flow will be iterated as errors being uncovered.

5.1 System-level Verification

As a first step, VHDL may be used to model and simulate aspects of the complete system containing one or more devices. This may be a fully functional description of the system allowing the FPGA/ASIC specification to be validated prior to commencing detailed design. Alternatively, this may be a partial description that abstracts certain properties of the system, such as a performance model to detect system performance bottle-necks[24].

5.2 RTL design and testbench creation

Once the overall system architecture and partitioning are stable, the detailed design of each FPGA/ASIC can commence[25]. This starts by capturing the design in VHDL at the register transfer level, and capturing a set of test cases in VHDL. These two tasks are complementary, and are sometimes performed by different design teams in isolation to ensure that the specification is correctly interpreted. The RTL VHDL should be synthesizable if automatic logic synthesis is to be used. Test case generation is a major task that

---

[21] *vt.* & *vi.* 伸展，伸长

[22] *vt.* 影响

[23] *n.* 排列，阵列

[24] *n.* 瓶颈

[25] *vt.* & *vi.* 开始，着手

requires a disciplined approach and much engineering ingenuity[26]: the quality of the final FPGA/ASIC depends on the coverage of these test cases.

5.3 RTL verification

The RTL VHDL is then simulated to validate the functionality against the specification. RTL simulation is usually one or two orders of magnitude faster than gate level simulation, and experience has shown that this speed-up is best exploited[27] by doing more simulation, not spending less time on simulation.

In practice it is common to spend 70%～80% of the design cycle writing and simulating VHDL at and above the register transfer level, and 20%～30% of the time synthesizing and verifying the gates.

5.4 Look-ahead Synthesis

Although some exploratory synthesis will be done early on in the design process, to provide accurate speed and area data to aid in the evaluation of architectural decisions and to check the engineer's understanding of how the VHDL will be synthesized, the main synthesis production run is deferred until functional simulation is complete. It is pointless to invest a lot of time and effort in synthesis[28] until the functionality of the design is validated.

6. Benefits of using VHDL

6.1 Executable specification

It is often reported that a large number of ASIC designs meet their specifications first time, but fail to work when plugged into a system. VHDL allows this issue to be addressed in two ways: A VHDL specification can be executed in order to achieve a high level of confidence in its correctness before commencing design, and may simulate one to two orders of magnitude[29] faster than a gate level description. A VHDL specification for a part can form the basis for a simulation model to verify the operation of the part in the wider system context (e.g. printed circuit board simulation). This depends on how accurately the specification handles aspects such as timing and initialization.

---

[26] *n.* 机灵，独创性，精巧，灵活性，聪明才智

[27] *vt.* 开发

[28] *n.* 综合，合成

[29] *n.* 数量，巨大，广大，量级

Behavioural simulation can reduce design time by allowing design problems to be detected early on, avoiding the need to rework designs at gate level. Behavioural simulation also permits design optimization by exploring alternative architectures, resulting in better designs.

6.2 Tools

VHDL descriptions of hardware design and test benches are portable between design tools, and portable between design centres and project partners. You can safely invest in VHDL modelling effort and training, knowing that you will not be tied in to a single tool vendor, but will be free to preserve your investment across tools and platforms. Also, the design automation tool vendors are themselves making a large investment in VHDL, ensuring a continuing supply of state-of-the-art VHDL tools.

6.3 Technology

VHDL permits technology independent design through support for top down design and logic synthesis. To move a design to a new technology you need not start from scratch or reverse, engineer a specification, instead you go back up the design tree to a behavioural VHDL description, then implement that in the new technology knowing that the correct functionality will be preserved.

6.4 Benefits

(1) Executable specification.
(2) Validate spec[30] in system context (subcontract).
(3) Functionality separated from implementation.
(4) Simulate early and fast (manage complexity).
(5) Explore design alternatives.
(6) Get feedback (produce better designs).
(7) Automatic synthesis and test generation (ATPG for ASICs).
(8) Increase productivity (shorten[31] time-to-market).
(9) Technology and tool independence (though FPGA features may be unexploited).

[30] *n.* 说明，规格

[31] *vt. & vi.* 缩短，（使）变短

参 考 译 文

可编程逻辑控制器（PLC）

1. 什么是 PLC？

PLC 代表可编程逻辑控制器。PLC 是专门设计用于在恶劣工业环境（如极端温度、潮湿、干燥和/或多尘条件）下能够可靠运行的计算机。PLC 几乎应用于每个行业，包括制造、食品加工、汽车、石油和天然气及许多其他行业。

2. PLC 如何工作？

PLC 的工作可以很容易地理解为一种循环扫描方式，称为扫描周期。

PLC 扫描过程包括以下步骤：

（1）操作系统开始循环并监测时间。

（2）CPU 开始从输入模块读取数据并检查所有输入的状态。

（3）CPU 开始执行用户或应用程序，这些程序由继电器梯形逻辑或任何其他 PLC 编程语言编写。

（4）CPU 执行所有内部诊断和通信任务。

（5）根据程序结果，将数据写入输出模块，从而更新所有输出。

只要 PLC 处于运行模式，这个过程就会持续进行。

3. PLC 的物理结构

PLC 的结构几乎类似于计算机的体系结构。PLC 持续监控来自各种输入传感设备（如加速度计、体重秤、硬连线信号等）的输入值，并根据生产和行业的性质产生相应的输出。典型的 PLC 由五部分组成，即机架或机箱、电源模块、CPU 模块、输入/输出模块及通信接口模块。

3.1 机架或机箱

在所有 PLC 系统中，PLC 机架或机箱是最重要的模块，它充当系统的主干。PLC 有不同的形状和尺寸。控制系统越复杂，需要的 PLC 机架或机箱就更大。

3.2 电源模块

电源模块用于为整个 PLC 系统提供所需的电能。它将可用的交流电转换为 CPU 模块和输入/输出模块所需的直流电。PLC 一般使用 24V 直流电。

3.3 CPU 模块

CPU 模块有中央处理器、ROM 和 RAM。ROM 用于存储操作系统、驱动程序和应用程序。RAM 用于存储程序和数据。CPU 是 PLC 的"大脑",具有微处理器,它取代了定时器、继电器和计数器。CPU 从传感器读取输入数据,进行处理,最后向控制设备发送命令。CPU 还包含其他电气部件,用于连接其他单元使用的电缆。

3.4 输入/输出模块

PLC 有一个输入/输出模块,专门作为输入和输出的接口。

输入设备可以是启动和停止按钮、开关等,输出设备可以是电加热器、阀门、继电器等。输入/输出模块有助于将输入设备和输出设备与微处理器连接起来。

输入模块有两个主要部分,即电源部分和逻辑部分。这两部分彼此电气隔离。

PLC 的输出模块的工作原理与输入模块类似,但过程相反。

3.5 通信接口模块

为了在 CPU 和通信网络之间传输信息,使用了智能输入/输出模块。这些通信模块有助于与放置在远程位置的其他 PLC 和计算机连接。

4. PLC 的种类

4.1 迷你 PLC

它们是小型、低成本控制器,非常适合简单的控制应用。它们通常比较大的控制器具有更少的输入/输出点,并且可以使用梯形逻辑或其他编程语言进行编程。迷你 PLC 由于尺寸小而可以快速安装,并且通常具有内置输入/输出功能(如数字输入、模拟输出和脉冲输出)。

4.2 模块化 PLC

它们由一个基本单元组成,该单元包含处理器模块和通信端口,以及可以添加以扩展系统功能的较小模块(见图 9-1)。模块化系统比固定系统提供了更大的灵活性,因为它们允许用户混合不同类型的输入/输出模块以满足其特定的应用要求。

(图略)

4.3 固定 PLC

它们专为特定任务而设计,一旦安装就无法轻易修改。然而,它们为许多重复性任务提供了经济高效的解决方案。固定系统适用于简单的过程控制应用,在操作过程中不需要频繁或快速改变参数。

4.4 微型 PLC

它们的复杂性介于迷你 PLC 和模块化 PLC 之间。它们通常是紧凑的设备,能够同时控制多个进程,而不需要额外的硬件组件(如某些模块化模型中的扩展卡或机架单元)(见图 9-2)。微控制器还可以集成通信功能(如以太网网络协议),以便轻松地集成到分布式自动化系统架构中。

(图略)

4.5 纳米 PLC

它们代表了最新一代的 PLC。这些超紧凑的设备使用了先进的微控制器,与专门的编程软件工具相结合,以降低成本,同时保持处理速度快且准确性高。纳米 PLC 甚至可用于涉及多轴运动或复杂机器视觉操作(如对象识别算法和模式匹配技术)的高度复杂应用。

4.6 安全 PLC

安全 PLC 用于实现工业自动化中的安全功能(见图 9-3)。它遵循国际安全标准,确保对人员、设备和环境的保护。安全 PLC 包含冗余和诊断功能,并具有安全完整性级别(SIL),以确保高可靠性和容错性。

(图略)

5. PLC 输出类型

5.1 模拟输出

为了控制需要不同程度控制的设备,人们使用模拟输出以产生连续的电压或电流信号。模拟输出通过改变它们产生的电压或电流水平来提供精确的控制。

5.2 继电器输出

PLC 和外部设备之间的电气隔离由 PLC 中的继电器输出提供。它们用于控制需要更大电流或更高电压的机械,如大型电机、接触器或功率继电器。

5.3 晶体管输出

晶体管输出指在 PLC 中使用晶体管作为开关器件的输出模块。与数字输出相比,晶体管输出可以处理更大的电流和更高的电压,这使其适合控制需要更大功率的设备。

6. PLC 编程

PLC 程序由一组文本或图形形式的指令组成,该程序表示管理 PLC 控制的过程的逻辑。PLC 编程语言主要有两大类,并进一步分为许多子分类类型。

(1)文本语言:指令表、结构化文本。
(2)图形形式:梯形图(LD)(即梯形逻辑)、功能块图(FBD)、顺序功能图(SFC)。

尽管这些 PLC 编程语言都可用于对 PLC 进行编程,但图形语言(如梯形逻辑)通常优于文本语言(如结构化文本编程)。

6.1 梯形逻辑

梯形逻辑是 PLC 编程中最简单的形式。它也被称为"中继逻辑"。继电器控制系统中使用的继电器触点使用梯形逻辑表示。

6.2 FBD

FBD 是一种在 PLC 中对多种功能进行编程的简单图形方法。功能块是一种程序指令单元，执行时会产生一个或多个输出值。

它由如图 9-4 所示的块表示。它可以用一个矩形块表示，左侧是输入，右侧是输出。它给出了输入状态和输出状态之间的关系。

（图略）

使用 FBD 的优点是在功能块上可以使用任意数量的输入和输出。当使用多个输入和输出时，可以将一个功能块的输出连接到另一个功能块的输入，从而构建 FBD（见图 9-5）。

（图略）

6.3 结构化文本编程

结构化文本是一种文本编程语言，它利用语句来确定要执行的内容。它遵循更传统的编程协议，但不区分大小写。

结构化文本具有一组表达式、赋值语句、循环语句、条件语句和函数。程序员可以用结构化文本语言编写可重用的代码块，并在不同的程序中使用这些代码块。结构化文本的好处是具有灵活性、可重用性、可读性、自记录性和可移植性。

Unit 10

Text A

Applications of Artificial Intelligence in Electrical Engineering

1. Introduction to Artificial Intelligence

Artificial intelligence (AI) is an area of computer science which is focused on developing technologies that can work intelligently and autonomously. AI utilizes advanced algorithms to enable computers to undertake tasks which typically require human-level awareness, such as making decisions, solving problems and understanding the environment. It has a wide range of applications in electrical engineering, ranging from power systems automation to industrial machine learning solutions. In this article, we'll explore how AI is being applied in electrical engineering and look at some examples of current projects using these technologies.

2. Overview of AI and Its Uses

AI is a rapidly advancing field of technology that has enabled machines to act in an increasingly autonomous and intelligent manner. AI-powered systems are being used in electrical engineering for a variety of applications, ranging from simulation and modelling to robotic design and process optimization. AI can be used to autonomously control electric power grids and other large electrical facilities like substations, and to improve the efficiency of managing complex manufacturing processes. It can also be utilized for predictive maintenance strategies within industrial plants. Additionally, it enables smart buildings with improved energy management capabilities while helping reduce manual errors through automation. In conclusion, AI holds much potential in revolutionizing the entire electrical engineering industry and its expansive range of uses allows organizations to leverage machine learning algorithms to optimize their operations across various domains.

3. AI in Electrical Engineering

AI is gaining traction in many different industries, including electrical engineering. It can be used to assist in tasks such as automating energy forecasting and optimizing the grid system for better efficiency. AI systems are designed to work with existing infrastructure and data acquired from sensors to build predictive models and make decisions, thereby improving performance, saving costs, and utilizing resources more effectively. Furthermore, AI algorithms can be used to automate aspects of product development, such as computer aided design (CAD), computer aided manufacturing (CAM), simulations and analytics on big datasets. Additionally, power electronics engineers utilize specialized reinforcement learning architectures called Hebbian networks, which operate efficiently when faced with unstable computing environments. Using this architecture can help in the design of high-voltage power lines or the development of intelligent cars or drones.

4. Benefits and Challenges of AI in Electrical Engineering

The use of AI in electrical engineering offers a number of advantages. Through AI, engineers can leverage sophisticated models and techniques to find new solutions and optimize existing ones. Furthermore, the use of machine learning algorithms within electrical engineering can automate many tedious tasks that are often laborious for humans to do by hand. Automation also brings potential cost-savings as different processes are streamlined and more efficient designs are found in shorter periods of time, thus decreasing labour costs.

Despite the various benefits associated with incorporating AI into electrical engineering projects, there remain some challenges before this technology is widely adopted. One issue is data availability. While advances have been made in sensor technologies to provide real-time datasets they still remain largely expensive and unreliable, making them difficult for smaller organizations working on constrained budgets to make full use of what AI has to offer. In addition, significant barriers exist around interpretability due to recent progress being focused on training autonomous machines. These machines may be able to interpret complex data but are unable yield meaningful explanations when decisions are reached. This means that decision accuracy remains harder for those who are not knowledgeable about underlying technologies behind deep learning algorithms. Therefore, leading teams at times mistrust results obtained through automated systems. Finally, it is also necessary to consider the "black box" issue, where there may be inaccurate correlations between inputs and outputs that have not been discovered, but if the assumptions proposed by these systems are considered facts rather than estimates, they may lead to harmful consequences[1].

5. Automating Power Grid Management with AI

AI has great potential to revolutionize the electrical engineering space. By using AI technology, power grid management can be automated to provide an efficient and reliable supply of energy with improved security. AI technologies such as machine learning can be used to detect anomalies in the power system by recognizing patterns from past data that may not be noticed otherwise. This automation can allow for quicker response times when it comes to problems related with power grid status due to predictive analytics capabilities. Furthermore, AI systems may also be leveraged for better demand forecasting and peak load balancing in order to ensure a stable electricity supply even during high usage periods. Automating power grids with AI technology can improve the efficiency and reliability of the entire country's energy supply at a macro level through more intelligent decision-making processes supported by intelligent algorithms running in the background[2].

6. Using AI to Develop Smart Electrical Systems

AI is increasingly being used to develop smart electrical systems in a wide variety of industries and applications. AI technology can be leveraged to create automated, adaptive and intelligent control systems for various types of devices, from actuators controlling motorized valves and pumps, to digital motors controlling machines like robots. Additionally, AI algorithms can provide predictive maintenance capabilities for major components such as transformers or circuit breakers in order to improve system reliability. This results in reduced downtime and downtime-related costs while eliminating the need for manual inspections without sacrificing safety standards. Furthermore, it helps industrial engineers increase their knowledge base by offering feedback on complex operations that aids predictive decision making based on historical data points collected over time[3]. Ultimately, using AI enables more efficient resource usage while keeping production output high with much higher accuracy than traditional methods ever could have achieved.

7. AI for Automated Quality Control in Electrical Engineering

Using AI for automated quality control in electrical engineering has become increasingly popular due to its ability to detect as well as recognize complex patterns quickly and accurately. AI-driven systems can rapidly help detect discrepancies and inconsistencies between data structures and expected results in processes such as motor control, circuit design, and embedded system development. With accurate analytics of the quality control process based on AI algorithms, optimal objectives along with parameters associated with them can be configured, which significantly reduce time taken for testing electrical engineering devices and circuits while also eliminating errors related to manual operations[4]. Moreover, automation helps track compliance measurements more precisely, resulting better overall product quality and durability.

8. Applications of AI in Diagnostics in Electrical Engineering

AI can significantly improve accuracy and efficacy in diagnostics for electrical engineering equipment. AI-powered detection and electronic monitoring systems enable preventive maintenance of sophisticated electronics. The algorithms employed by these systems allow the systems to detect anomalies, recognize factors such as ambient temperature or voltage variation, interpret data retrieved from sensors or other data sources quickly, and accurately calculate the root cause analysis (RCA) with minimal intervention from engineers. By leveraging AI-driven predictive analytics, preventative maintenance plans can be created to proactively anticipate potential problems before they arise. This ensures optimum operational performance while cutting down on costs associated with emergency repairs due to unforeseen issues that may occur over time.

9. Machine Learning for Maintenance of Electrical Systems

Machine learning is rapidly growing as a beneficial technology to effectively maintain electrical systems. Breakdowns of critical components can be predicted with high accuracy by using machine learning algorithms, enabling engineers and technicians to make programmed maintenance plans that save time and money while ensuring optimal efficiency of the system. These predictions also provide insights into potential weak points in the system before they cause any disruption or costly repairs, thus minimizing downtime and negative side effects from production losses or environmental damage. Machine learning techniques such as clustering, network analysis and pattern recognition are especially well-suited for modeling electrical data sources like fault codes and oscilloscope readings in order to accurately monitor the components of an electrical system continuously over its lifespan.

10. Conclusion

AI has become increasingly popular and applicable in the field of electrical engineering due to its potential for optimizing solutions, automating tasks, and solving complex problems. AI technology can be used in such applications as renewable energy projects, smart grid automation, fault detection and prevention systems, drone-based inspection operations, neural network based circuit design, etc. Companies utilizing AI may benefit from significant cost reductions from automation and still deliver high quality end products thanks to advanced optimization capabilities. Ultimately this could result in an increase in competitive advantage across many enterprises in the electrical engineering industry.

New Words

autonomously	[ɔːˈtɒnəməslɪ]	adv.	自主地
undertake	[ˌʌndəˈteɪk]	vt.	承担，从事；保证；同意
awareness	[əˈweənəs]	n.	意识
environment	[ɪnˈvaɪrənmənt]	n.	环境，外界；周围
substation	[ˈsʌbsteɪʃn]	n.	变电站，变电所
predictive	[prɪˈdɪktɪv]	adj.	预测性的
plant	[plɑːnt]	n.	工厂
potential	[pəˈtenʃl]	adj.	潜在的，有可能的
		n.	潜力，潜能；电位，势能
leverage	[ˈliːvərɪdʒ]	v.	利用
		n.	杠杆作用；优势，力量
optimize	[ˈɒptɪmaɪz]	vt.	使最优化
traction	[ˈtrækʃn]	n.	吸引力，牵引力
dataset	[ˈdeɪtəset]	n.	数据集
network	[ˈnetwɜːk]	n.	网络
unstable	[ʌnˈsteɪbl]	adj.	不稳固的，易变的

drone	[drəʊn]	n. 无人机
tedious	[ˈtiːdɪəs]	adj. 单调沉闷的；冗长乏味的；令人生厌的
streamline	[ˈstriːmlaɪn]	vt. 使简单化
unreliable	[ˌʌnrɪˈlaɪəbl]	adj. 不可靠的，不能信任的
interpretability	[ɪntɜːprɪˈtəbɪlɪtɪ]	n. 可解释性
progress	[ˈprəʊgres]	n. 进程；进步，前进
meaningful	[ˈmiːnɪŋfl]	adj. 有意义的，有意图的
explanation	[ˌekspləˈneɪʃn]	n. 解释；说明
knowledgeable	[ˈnɒlɪdʒəbl]	adj. 有知识的，有见识的
mistrust	[ˌmɪsˈtrʌst]	v. 不信任，怀疑
		n. 不信任
correlation	[ˌkɒrəˈleɪʃn]	n. 相关性
assumption	[əˈsʌmpʃn]	n. 假定，假设
consequence	[ˈkɒnsɪkwəns]	n. 结果；后果
detect	[dɪˈtekt]	vt. 检测；[电子学]检波
anomaly	[əˈnɒməlɪ]	n. 异常，反常
pattern	[ˈpætn]	n. 模式
status	[ˈsteɪtəs]	n. 状态
background	[ˈbækgraʊnd]	n. 后台；背景
adaptive	[əˈdæptɪv]	adj. 适应的，有适应能力的
actuator	[ˈæktʃʊeɪtə]	n. 执行器，执行机构；激励者
robot	[ˈrəʊbɒt]	n. 机器人
transformer	[trænsˈfɔːmə]	n. 变压器
downtime	[ˈdaʊntaɪm]	n. 停工期
eliminate	[ɪˈlɪmɪneɪt]	vt. 排除，消除
discrepancy	[dɪsˈkrepənsɪ]	n. 差异；不符合；不一致
inconsistency	[ˌɪnkənˈsɪstənsɪ]	n. 不一致，不协调
configure	[kənˈfɪgə]	v. 配置；设定
durability	[ˌdjʊərəˈbɪlətɪ]	n. 耐久性，持久性
diagnostic	[ˌdaɪəgˈnɒstɪk]	adj. 诊断的，判断的
		n. 诊断法，诊断程式
anticipate	[ænˈtɪsɪpeɪt]	vt. 预感；预见；预料
optimum	[ˈɒptɪməm]	adj. 最适宜的
		n. 最佳效果
unforeseen	[ˌʌnfɔːˈsiːn]	adj. 未预见到的；意外的，偶然的
beneficial	[ˌbenɪˈfɪʃl]	adj. 有利的，有益的
breakdown	[ˈbreɪkdaʊn]	n. 损坏，故障
insight	[ˈɪnsaɪt]	n. 洞察力；见解；直觉
cluster	[ˈklʌstə]	v. 聚集

		n. 丛；簇；群
oscilloscope	[əˈsɪləskəʊp]	n. 示波器
lifespan	[ˈlaɪfspæn]	n. 寿命；存在期；使用期；有效期
applicable	[əˈplɪkəbl]	adj. 适当的；可应用的

Phrases

computer science	计算机科学
be focused on	专注于……
make decision	决策
power systems automation	电力系统自动化
AI-powered system	基于人工智能的系统，人工智能驱动的系统
be utilized for	用于
smart building	智能建筑，智能楼宇
manual error	手动错误
predictive model	预测模型
save cost	节省成本
reinforcement learning architecture	强化学习结构
high-voltage power line	高压电力线路
intelligent car	智能汽车
a number of	许多，一些
autonomous machine	自动机器
deep learning	深度学习
black box	黑盒
response time	响应时间
be leveraged for	被利用
peak load	最大负荷，峰值负荷
usage period	使用期
macro level	宏观层面
smart electrical system	智能电气系统
intelligent control system	智能控制系统
digital motor	数字电动机，数字马达
circuit breaker	断路开关，断路器
in order to	为了……
manual inspection	手动检测，人工检验
historical data	历史数据
traditional method	传统方法
quality control	质量管理
data structure	数据结构
circuit design	电路设计

embedded system	嵌入式系统
electronic monitoring system	电子监控系统
ambient temperature	环境温度，背景温度
voltage variation	电压变化
preventative maintenance plan	预防性维护计划
cut down on cost	削减成本
emergency repair	应急维修
critical component	关键部件
pattern recognition	模式识别
fault code	故障码
renewable energy project	可再生能源项目
fault detection and prevention system	故障检测与预防系统
drone-based inspection operation	基于无人机的检查操作
neural network	神经网络

Abbreviations

AI (Artificial Intelligence)	人工智能
CAM (Computer Aided Manufacturing)	计算机辅助制造
RCA (Root Cause Analysis)	根本原因分析

Notes

[1] Finally, it is also necessary to consider the "black box" issue, where there may be inaccurate correlations between inputs and outputs that have not been discovered, but if the assumptions proposed by these systems are considered facts rather than estimates, they may lead to harmful consequences.

本句是由 but 连接的两个并列句。在第一个句子中，it 是形式主语，真正的主语是动词不定式短语 to consider the "black box" issue。where there may be inaccurate correlations between inputs and outputs that have not been discovered 是一个非限定性定语从句，对 the "black box" issue 进行补充说明。在该从句中，that have not been discovered 是一个定语从句，修饰和限定 inaccurate correlations。在 but 后面的句子中，if the assumptions proposed by these systems are considered facts rather than estimates 是一个条件状语从句，修饰谓语 may lead to。在该从句中，proposed by these systems 是一个过去分词短语，作定语，修饰和限定 the assumptions。rather than 的意思是"而不是；而不"，lead to 的意思是"导致"。

本句意为：最后，还需要考虑"黑盒"问题，其中输入和输出之间可能存在未被发现的不准确的相关性，但如果这些系统提出的假设被视为事实而不是估计，则可能会导致有害的后果。

[2] Automating power grids with AI technology can improve the efficiency and reliability of the entire country's energy supply at a macro level through more intelligent decision-making processes supported by intelligent algorithms running in the background.

本句中，Automating power grids with AI technology 是一个动名词短语，作主语。through more intelligent decision-making processes supported by intelligent algorithms running in the background 是一个介词短语，作方式状语，修饰谓语 can improve。在该方式状语中，supported by intelligent algorithms running in the background 是一个过去分词短语，作定语，修饰和限定 more intelligent decision-making processes。在该过去分词短语中，running in the background 是一个现在分词短语，作定语，修饰和限定 intelligent algorithms。at a macro level 也是一个状语，意思是"在宏观层面"。

本句意为：利用人工智能技术可以实现电网自动化，它通过后台运行的智能算法实现更加智能的决策过程，这样可在宏观层面提高整个国家能源供应的效率和可靠性。

[3] Furthermore, it helps industrial engineers increase their knowledge base by offering feedback on complex operations that aids predictive decision making based on historical data points collected over time.

本句中，by offering feedback on complex operations that aids predictive decision making based on historical data points collected over time 是一个介词短语，作方式状语，修饰谓语 helps industrial engineers increase their knowledge base。在该方式状语中，that aids predictive decision making based on historical data points collected over time 是一个定语从句，修饰和限定 feedback。collected over time 是一个过去分词短语，作定语，修饰和限定 historical data points。based on 的意思是"基于，根据"。

本句意为：此外，它通过提供复杂操作的反馈来帮助行业工程师增加他们的知识库，这些反馈有助于根据长期收集的历史数据点做出预测性决策。

[4] With accurate analytics of the quality control process based on AI algorithms, optimal objectives along with parameters associated with them can be configured, which significantly reduce time taken for testing electrical engineering devices and circuits while also eliminating errors related to manual operations.

本句中，which significantly reduce time taken for testing electrical engineering devices and circuits while also eliminating errors related to manual operations 是一个非限定性定语从句，对其前面的句子进行补充说明。在该从句中，taken for testing electrical engineering devices and circuits 是一个过去分词短语，作定语，修饰和限定 time。related to manual operations 也是一个过去分词短语，作定语，修饰和限定 errors。

本句意为：基于人工智能算法对质量控制过程进行准确分析，可以配置最佳目标及其相关参数，这大大减少了测试电气工程设备和电路所需的时间，同时也消除了与手动操作相关的错误。

Exercises

【Ex.1】根据课文内容，回答以下问题。

1. What does AI stand for? What is it?

2. What can AI algorithms be used to?

3. What are the challenges of AI in electrical engineering?

4. Why has using AI for automated quality control in electrical engineering become increasingly popular?

5. Why has AI become increasingly popular and applicable in the field of electrical engineering?

【Ex.2】把下面的单词填写到合适的空格中。

| A | microcontroller | B | power | C | transistors | D | chip | E | integrated |
| F | disadvantage | G | material | H | discrete | I | smaller | J | memory |

What Is an Integrated Circuit?

An integrated circuit (IC) is a semiconductor wafer on which thousands or millions of small resistors, capacitors, diodes, and transistors are created. Computer ___1___, oscillator, counter, amplification, logic gate, timer, ___2___, or processor are all examples of integrated circuits. All modern electrical gadgets have an IC as their basic building element. It's an ___3___ system of several miniaturized and interconnected components embedded in a thin silicon chip.

A huge number of tiny MOSFETs (Metal-Oxide-Semiconductor Field Effect Transistors) are packed on a small ___4___ and coupled to form an integrated circuit. This produces circuits that are significantly faster, ___5___, and less costly than discrete circuits built with discrete electronic components.

The electronics industry has hurried to adopt standardized ICs in designs employing discrete ___6___ due to mass production capability, reliability, and the building-block approach to integrated circuit design. ICs have two key advantages over ___7___ circuits: performance and cost.

Because the components inside an IC have quicker switch times and consume less ___8___ due to their compact size and proximity, performance is significantly higher in ICs than in discrete counterparts.

Because ICs are manufactured as a unit using photolithography rather than being built one transistor at a time, the cost is very low. Packaged circuits also utilize far less ___9___ than discrete circuits.

However, ICs have a significant ___10___: the high expenses of designing them and creating the photolithography masks. As a result, ICs are only financially viable when enormous manufacturing volumes are expected, allowing profit margins to justify them.

【Ex.3】把下列句子翻译成中文。

1. Substation automation system uses memory database for real-time data access.

2. In order to optimize the process, formulation of objective function is necessary.

3. The company has identified fifty potential customers at home and abroad.

4. A firewall provides an essential security blanket for your computer network.

5. A dataset is an ordered collection of data.

6. You can track test execution progress efficiently and easily.

7. Their instruments detected very faint radio waves at a frequency of 3 kilohertz.

8. People alter their voices in relationship to background noise.

9. A transformer is a passive component that transfers electrical energy from one electrical circuit to another circuit, or multiple circuits.

10. An oscilloscope is a type of electronic test instrument that graphically displays varying voltages of one or more signals as a function of time.

【Ex.4】把下列短文翻译成中文。

Deep learning is a subfield of machine learning that involves the use of neural networks to model and solve complex problems. Neural networks are modeled after the structure and function of the human brain and consist of layers of interconnected nodes that process and transform data.

The key characteristic of deep learning is the use of deep neural networks, which have multiple layers of interconnected nodes. These networks can learn complex representations of data by discovering hierarchical patterns and features in the data. Deep learning algorithms can automatically learn and improve from data.

Deep learning has achieved significant success in various fields, including image recognition, natural language processing, speech recognition, and recommendation systems. Some of the popular deep learning architectures include convolutional neural networks (CNNs), recurrent neural networks (RNNs), and deep belief networks (DBNs).

Training deep neural networks typically requires a large amount of data and computational resources. However, the availability of cloud computing and the development of specialized hardware, such as graphics processing units (GPUs), has made it easier to train deep neural networks.

【Ex.5】通过 Internet 查找资料，借助电子词典和辅助翻译软件，完成以下技术报告。通过 E-mail 发送给老师，并附上收集资料的网址。
1. 人工智能在电气工程领域的应用并列出它们的主要技术特点。
2. 大数据在电气工程领域的应用并列出它们的主要技术特点。

Text B

Top Skills for an Electrical Engineer

1. 5 Skills Essential for an Electrical Engineer

Electrical engineering is a dynamic field requiring diverse skills. The following are the top 5 skills essential for an electrical engineer.

1.1 Technical skills

As an electrical engineer, you must have a firm grasp of technical skills to design, develop and maintain electrical systems. Some examples of essential technical skills for electrical engineers include:

Circuit design and analysis—ability to design and analyze analog, digital, and mixed-signal circuits.

Control systems—ability to design, model, and optimize control systems, including feedback and open-loop control systems. These skills for an electrical engineer are instrumental in controlling and regulating various systems, including robotics, automation, and power systems.

Digital signal processing—ability to design and optimize digital systems, such as communication, audio and video processing, and biomedical systems.

Electromagnetics—ability to design, develop, and optimize electromagnetic systems, including antennas, radar, and wireless communication systems.

Electronics—ability to design, develop, and optimize electronic circuits for various applications, including power electronics, signal processing, and control systems.

Power systems—ability to design, develop, and optimize power systems for various applications, including renewable energy systems, power distribution systems, and electric vehicles.

Programming languages—proficiency in programming languages, including C++, Python, and MATLAB. These skills help create software for embedded systems, control systems, and other applications.

Simulation software—familiarity with simulation software, including SPICE, PSIM, and PSCAD. These skills for an electrical engineer can help simulate and analyze electrical systems and circuits.

1.2 Analytical skills

Electrical engineers need to be able to analyze complex systems and data to make informed decisions. Some examples of critical analytical skills for electrical engineers include:

Data analysis—ability to identify trends, patterns, and anomalies in large datasets. They use this skill to analyze data from various sources, including sensors, meters, and other devices.

Fault analysis—ability to identify and diagnose faults in electrical systems. They use this skill to troubleshoot and repair faulty electrical systems and components.

Optimization—ability to find the best solution to a problem by maximizing or minimizing a specific objective function.

Risk analysis—ability to identify and assess potential electrical systems and components risks. They use this skill to develop and implement risk mitigation strategies.

Statistical analysis—ability to analyze and interpret data using statistical methods, including regression analysis, hypothesis testing, and analysis of variance.

Systems analysis—ability to identify and analyze a system's components, inputs, outputs, and interactions. These skills can help understand and optimize complex systems, including power, communication, and control systems.

1.3 Communication skills

Besides technical and analytical skills, practical communication skills are critical for electrical engineers to succeed. As an electrical engineer, you must communicate complex technical concepts to non-technical stakeholders such as clients, managers, and colleagues. Here are some specific communication skills that are most important for electrical engineers:

Technical writing—ability to create technical documentation such as design specifications, user manuals, and test reports. It's essential to communicate technical information clearly and concisely in writing.

Presentation skills—ability to present their work to clients, colleagues, or management. Practical presentation skills for an electrical engineer are crucial to deliver a compelling message and convincing stakeholders of the value of your work.

Project management—ability to work on projects with teams of professionals, including other engineers, technicians, project managers, and all key project stakeholders. Communication skills are essential to coordinate project activities, ensure everyone is on the same page, and achieve project goals.

Team collaboration—ability to work as a well-collaborated team member on large projects. It's essential to work collaboratively with other professionals, communicate ideas effectively, and resolve conflicts constructively.

Verbal communication—ability to communicate ideas and technical information effectively in person or via video conferencing.

1.4　Leadership skills

Leadership skills are also necessary for electrical engineers in supervisory or managerial positions. Even if you don't have direct reports, leadership skills for electrical engineers can help them be more effective team members and contributors. Here are some critical leadership skills for electrical engineers:

Decision-making—ability to make decisions based on incomplete or ambiguous information. Practical decision-making skills are crucial to make informed choices that align with project goals and timelines.

Project management—ability to take ownership and responsibility for managing projects from start to finish. This requires planning, organizing, and coordinating project activities, managing project risks, and ensuring that projects are delivered on time and within budget.

Strategic planning—responsible for developing long-term strategic plans for their organizations. This requires analyzing industry trends, assessing organizational strengths and weaknesses, and developing plans that align with company goals.

Team management—managing teams of engineers, technicians, or other professionals. This requires inspiring and motivating team members, providing feedback and coaching, and ensuring they have the necessary resources to succeed.

Time management—ability to work on multiple projects simultaneously. Practical time management skills for an electrical engineer are essential to ensure that you can prioritize tasks, meet deadlines, and deliver high-quality work on time.

1.5　Problem-solving skills

In this section, we will discuss the problem-solving skills which are most important for electrical engineers. Some examples might include:

Critical thinking—ability to analyze data, evaluate information, and consider different perspectives to make sound decisions.

Innovative thinking—ability to craft new and creative ideas to solve complex problems in the field.

Troubleshooting—ability to identify and solve problems with electrical systems, equipment, and machinery quickly and efficiently.

Root cause analysis—ability to identify the underlying causes of problems and develop solutions that address those causes.

Systematic approach—trained for problem-solving that requires gathering information, analyzing data, and developing and testing solutions.

Attention to detail—ability to pay attention to details and all aspects of complex electrical systems, circuits, and equipment to avoid errors and ensure safety.

2. Electrical Engineering Skills Listed on Resumes

Here are the top 5 electrical engineering skills that must be present on your resume.

2.1 Project management

Project management skill for an electrical engineer is in high demand in today's job market. Especially for those of you who want to earn a little higher or more than the average salary and eventually become a team leader, this skill is a must.

Employers look for this skill in your resume to ensure that you will become a part of the succession pipeline and have the required competency to lead a team in the future and have a supervising outlook.

2.2 AutoDesk AutoCAD

You must know how to use AutoCAD and deliver amazing visuals for your electrical engineering ideas using this software. This computer-aided design tool not only brings your ideas to reality but also makes it easier for decision-makers to understand your core concept. You must have a command over this if you want to impress your managers and make an impact on your team.

2.3 Engineering design

As an electrical engineer, having a design background is essential. Not only should this be represented on your resume, but it's vital that you demonstrate mastery of the skill as well.

Suppose you need to be more competent in engineering design or the seven steps involved in the engineering design process, starting from the definition of the idea itself and right down to prototype development and its testing and improvements. In that case, your chances of securing a job in today's competitive market could be greater.

2.4 MATLAB

Another essential skill for any electrical engineer is using and understanding MATLAB.

MATLA enables you to use your algorithms and perform mathematical operations smoothly.

2.5　Programmable logic controllers/automation

As automation and robotics rapidly become more pervasive throughout various industries, employers are increasingly seeking out individuals with expertise in programmable logic controllers (PLCs) or other relevant forms of technology to help streamline the process. These tech-savvy pros can automate procedures while saving time and money through reduced overhead costs, allowing businesses to stay ahead of the curve with advanced approaches.

New Words

firm	[fɜːm]	adj. 稳固的；强有力的
grasp	[grɑːsp]	v. & n. 掌握；理解
instrumental	[ˌɪnstrəˈmentl]	adj. 有帮助的；起作用的
biomedical	[ˌbaɪəʊˈmedɪkl]	adj. 生物医学的
electromagnetic	[ɪˌlektrəʊmæɡˈnetɪk]	adj. 电磁的
antenna	[ænˈtenə]	n. 天线
radar	[ˈreɪdɑː]	n. 雷达；无线电探测器
proficiency	[prəˈfɪʃnsɪ]	n. 熟练，精通
familiarity	[fəˌmɪliˈærətɪ]	n. 熟悉；通晓
troubleshoot	[ˈtrʌblʃuːt]	v. 检修
identify	[aɪˈdentɪfaɪ]	vt. 识别，认出；确定
assess	[əˈses]	v. 评估；估价；估算
mitigation	[ˌmɪtɪˈɡeɪʃn]	n. 缓解，减轻
interpret	[ɪnˈtɜːprət]	v. 解释；领会
hypothesis	[haɪˈpɒθəsɪs]	n. 假设，假说；[逻]前提
variance	[ˈveərɪəns]	n. 方差
stakeholder	[ˈsteɪkhəʊldə]	n. 股东；利益相关者
concisely	[kənˈsaɪslɪ]	adv. 简明地
presentation	[ˌpreznˈteɪʃn]	n. 表达；介绍
conflict	[ˈkɒnflɪkt]	n. 争执；冲突
	[kənˈflɪkt]	v. 冲突
innovative	[ˈɪnəveɪtɪv]	adj. 创新的；革新的
systematic	[ˌsɪstəˈmætɪk]	adj. 有系统的，有步骤的
competency	[ˈkɒmpɪtənsɪ]	n. 资格，能力
impress	[ɪmˈpres]	v. 给……深刻印象；使……意识到
demonstrate	[ˈdemənstreɪt]	v. 证明；说明
mastery	[ˈmɑːstərɪ]	n. 精通，熟练
improvement	[ɪmˈpruːvmənt]	n. 改善，改进
pervasive	[pəˈveɪsɪv]	adj. 普遍的；扩大的

tech-savvy	[tek ˈsævɪ]		*adj.* 懂技术的

Phrases

electrical engineer	电气工程师
circuit design and analysis	电路设计与分析
open-loop control system	开环控制系统
digital signal processing	数字信号处理
wireless communication system	无线通信系统
simulation software	仿真软件，模拟软件
be able to	能，会，能够
data analysis	数据分析
fault analysis	事故分析，故障分析
risk analysis	风险分析
statistical analysis	统计分析
regression analysis	回归分析
technical documentation	技术文档，技术资料
user manual	用户手册
test report	测试报告
project management	项目管理
project manager	项目经理，项目管理人
team collaboration	团队协作
video conferencing	视频会议
critical thinking	批判性思维
creative idea	创见，创意
core concept	核心概念
prototype development	原型开发
mathematical operation	数学运算
seek out	寻找
overhead cost	管理费用，间接成本

Exercises

【Ex.6】根据文章所提供的信息判断正误。

1. An electrical engineer must have a firm grasp of technical skills to design, develop and maintain electrical systems.
2. The first essential technical skill for electrical engineers mentioned in the passage is circuit design and analysis.
3. Electrical engineers do not have to be able to analyze complex systems and data to make informed decisions.
4. Electrical engineers should have the ability to identify and diagnose faults in electrical

systems.
5. Practical communication skills are not important for electrical engineers to succeed.
6. Practical presentation skills for an electrical engineer are crucial to deliver a compelling message except convincing stakeholders of the value of your work.
7. Strategic planning requires analyzing industry trends, assessing organizational strengths and weaknesses, and developing plans that align with company goals.
8. Electrical engineers should have critical thinking—ability to analyze data, evaluate information, and consider different perspectives to make sound decisions .
9. Project management skill for an electrical engineer is not in high demand in today's job market.
10. Not all electrical engineers should have the skill to use and understand MATLAB.

科技英语翻译知识

篇章翻译

以上几部分讨论的都是有关词汇和单句的翻译。在对这两部分的翻译有了初步的把握后，应该认识到人们翻译的不会是孤零零的单句，而是完整的语篇材料，或有明确的上下文的段落。所以，在翻译过程中，最终不能从一词、一句甚至一段出发，而应该从整个篇章出发。篇章是一个层次体系，它由段落组成，段落又由句子组成，句子又可分为小句，小句里又有词和词组。词与词、句与句、段落与段落之间都有形式上和内容上的联系。翻译策略体现在每个层次上都有侧重点，体现在段落和篇章层次上。应注意解决句子之间、段落之间的逻辑连接。因此，在翻译过程中，要注意整个语篇的翻译而不是单个句子的简单相加，要以篇章为基础。下面以两个例子做分析。

例一

The best projectiles for this purpose were found to be neutrons. Their mass enables them to pierce the shells of electrons and their small size and electrical neutrality enables them to penetrate the nucleus itself. Once inside the nucleus the intruding neutron may cause a restructuring and rearrangement of protons and neutrons without causing outward disturbance; in this case, the neutron is absorbed and an isotopic atom of the element is formed. But the intruding neutrons may alternatively disrupt the heavy nucleus, causing it to disintegrate into two or more parts which then become the atoms of two or more different elements; a transmutation of an element has occurred. When heavy atoms are split in this way some loss of mass occurs and this loss of mass is converted into an equivalent quantity of energy according to Einstein's law $E = mc^2$ where c is the velocity of light.

人们发现，用于此目的的发射体就是中子。中子的质量能够粉碎电子壳层，其极小的体积和不带电的特性可以使其自身穿透原子核。一旦进入原子核，"侵入"的中子便可引起质子与中子的重构和重排，且对外层无丝毫影响。此时，中子被吸收，一种新的同位素原子也就诞生了。但是,"侵入"的中子还可能使重原子核分裂,使其分裂成两个或更多部分,成为两种或多种完全不同的元素。原来的元素就发生了蜕变。当重原子以这种方式被分裂时,也就发生了质量损失。根据爱因斯坦定律 $E = mc^2$(式中 c 为光速),这种质量转化成了相等的能量。

分析:英语中为了避免重复,通常用代词指代已提到过的前文事物,代词也因此把两句话连接了起来。原文中第二句话的 their 指代第一句话中出现的 "neutrons",因此在翻译时,根据汉语习惯应该重复"中子"。第四句是由两个句子组成的长句子,用分号隔开。前一个分句是说"侵入"的中子可能使重原子核分裂成多个部分,成为不同的元素。后一个分句说原来的元素发生了蜕变。很显然后一个分句就是前一个分句的结果,因此在翻译中,要把这层关系表达出来,而不能孤零零地把第二个分句翻译成"原来的元素发生了蜕变"。

例二

According to growing body of evidence, the chemicals that make up many plastics may migrate out of the material and into foods and fluids, ending up in your body. Once there they could make you very sick indeed. That's what a group of environmental watchdogs has been saying, and the medical community is starting to listen.

一些环保检查人员指出,越来越多的证据表明,许多塑料制品的化学成分会移动到食物或流体上去,最终进入人体内。一旦进入人体,便会使人生病。这一问题也开始受到医学界的关注。

分析:原文第三句"That's what a group of environmental watchdogs has been saying"表示前两句话是谁的观点,所以可以根据汉语习惯把它调到第一句。第二句中的 there 指代的是第一句的 in your body,但是在翻译中,要把"人体"重复出来,这样译文才显得自然。listen 是"听"的意思,但是结合整个篇章可以知道,关于一个问题,人们给予的应是"关注",因此,这里的 listen 做了灵活处理。

Reading Material

阅读下列文章。

Text	Note
Types of Renewable Energy Resources 　　Renewable[1] energy is the energy that comes from the earth and has the ability to replenish[2] itself naturally. This kind of energy is sustainable[3], meaning it virtually never runs out because the sources it comes from replenish constantly. That being said, it can take some time to restore usable levels of renewable energy after depleting[4] a certain amount of it. 　　One of the greatest benefits of renewable energy sources is that they're much better for the environment than nonrenewable resources like gas and coal. Most of them produce no greenhouse gas[5]	[1] *adj.* 可再生的 [2] *vt.* 补充 [3] *adj.* 可持续的 [4] *v.* 耗尽,用尽 [5] greenhouse gas: 温室气体

emission[6], which is the primary driver of climate change. They can also reduce certain kinds of air and water pollution[7], which is better for our health and the health of our planet overall.

There are six major sources of renewable energy: solar, wind, hydroelectric[8], geothermal[9], biomass and ocean (or tidal[10]) energy. They're all inexhaustible[11] sources of energy because they rely on weather and other natural phenomena. Developing them throughout the world has the potential to create millions of jobs while saving the planet at the same time. In the long run, they're also much more cost-effective than using nonrenewable resources.

1. Solar Energy

Solar power[12] is a renewable energy resource that comes from the sun. It works by capturing the sun's energy with solar cells[13] on solar panels[14] and turning it into electricity or heat. When solar panels gather light from the sun and turn it into energy, they store this energy in batteries that people can use to power appliances and other systems in their homes.

Sunlight is functionally never-ending, so there's no limit to its potential for generating energy. Solar power does not produce carbon dioxide[15] or other air pollutants, making it very beneficial for the environment. In the long term, switching to solar energy also saves people money on their utility costs.

There are some limitations to solar energy, however. We have no control over the amount of sunlight that we receive or when we receive it. The sunlight a solar panel can receive depends on time, location, seasons and weather. Furthermore, absorbing[16] a useful quantity of sunlight takes a large surface area. Not everyone has space at home or in their yards to set up a network of solar panels that's large enough to produce a sufficient quantity of energy.

2. Wind Energy

Thanks to the fact that the sun heats the earth in an uneven[17] manner, we have wind. Wind power is a renewable energy source that we collect via turbines — machines that look like giant, modern windmills[18]. Turbines can reach the same height as skyscrapers[19], and the diameters of their blades[20] are almost as wide as one of these buildings, too. The blades spin when the wind hits them, which creates electricity by feeding the energy from the turning blades into

[6] *n.* 排放
[7] *n.* 污染
[8] *adj.* 水力发电的
[9] *adj.* 地热的
[10] *adj.* 潮水的，潮汐的
[11] *adj.* 无穷无尽的，用不完的

[12] solar power: 太阳能
[13] solar cell: 太阳能电池
[14] solar panel: 太阳能板

[15] carbon dioxide: 二氧化碳

[16] *v.* 吸收

[17] *adj.* 不均匀的
[18] *n.* 风车
[19] *n.* 摩天大楼
[20] *n.* 叶片

a generator. In windy locations, wind power can cost less than any other energy source.

Turbines are effective anywhere wind speeds are high. Open plains, hilltops and open water are ideal locations for these machines. Producing energy with turbines doesn't generate air pollution or carbon dioxide, making it a clean source of energy. The primary disadvantages of wind energy are inconsistency (many areas don't receive constant wind), noise disturbances[21] from the sounds the turbines make, the large amount of space the turbines take up and the fact that they can only be used in certain geographic locations.

[21] *n.* 打扰，困扰

3. Hydroelectric Energy

Hydroelectric power is generated by moving water. In the case of a dam[22], water runs through the structure's turbines to spin them, which creates energy for electricity. This is another clean energy source because it doesn't pollute the air.

[22] *n.* 水坝，水库

Among the top benefits of hydropower plants is their ability to accumulate[23] reservoirs[24] of energy for later use. This makes it possible to utilize less reliable renewable energy sources, such as solar and wind power, as the energy from hydroelectric power plants can be available as a backup when those others aren't.

[23] *v.* 积累
[24] *n.* 水库；蓄水池；储备，储藏，蓄积

To create a hydroelectric power plant, engineers have to dam a source of running water, such as a river. This can have negative effects on fish populations in the area, which can then affect the other animals that rely on those fish as a food source. Disrupting the food chain[25] can have a lot of negative outcomes. Additionally, if there is a drought[26], hydroelectric power plants can become less reliable because they need large quantities of water to function properly.

[25] food chain: 食物链
[26] *n.* 干旱

4. Geothermal Energy

The inner core of the earth is very hot. Because of this, heat continually rises up out of the earth's surface. The heat beneath the surface of the planet is called geothermal heat. The earth continuously produces this heat, which is what makes geothermal energy renewable.

To use geothermal energy, engineers drill wells several miles down into the earth, and hot water or steam rise up through these wells. The steam turns a turbine, which generates electricity, and the steam cools back into water so it can repeat this process again. The pollution created from collecting geothermal energy is minimal. Like hydropower, geothermal energy is a stable and virtually endless energy source.

The greatest disadvantage of geothermal energy is that it can only be used in places where drilling[27] deep into the earth is possible. In a lot of locations this isn't feasible, so engineers and companies have no access to this renewable energy resource. Geothermal power plants are often located in areas where there are lots of volcanoes[28], geysers[29] or hot springs because the heat is more accessible there. Also, geothermal energy can cause earthquakes[30], because digging the wells to access the energy changes the earth's structure and creates cracks.

[27] v. 钻（孔），打（眼）
[28] n. 火山
[29] n. 间歇喷泉
[30] n. 地震

5. Biomass Energy

Biomass is the material that comes naturally from plants and animals. It includes trees, plants, waste from crops[31], animal manure, human sewage and organic solid waste like cotton, paper, food, wool and wood scraps. We convert biomass into energy in various ways. This can include burning the materials to produce heat or using chemical, thermochemical[32] and biological conversion to produce different kinds of fuels. Combustion, or directly burning biomass, is the most common way to turn biomass into energy. Usually, this involves burning biomass to heat water and generate steam, which then creates electricity.

[31] n. 作物

[32] adj. 热化学的

Because humans, animals and plants are always creating waste, we'll likely never run out of biomass from which to produce energy. Plus, when we use waste, it doesn't go into a landfill[33]. However, there are some downsides to energy made from biomass: Burning these materials releases carbon dioxide, a greenhouse gas, into the air. It also pollutes the air with carbon monoxide[34]. What's more, when we clear land to grow crops specifically for the purpose of creating biomass, a lot of energy and space are wasted. Cutting down trees to use for biomass is harmful to the environment, too.

[33] n. 垃圾填埋场

[34] carbon monoxide: 一氧化碳

6. Ocean or Tidal Energy

Generating energy from the tides, waves and heat in the ocean is still a work in progress, but it can be done effectively in two ways. The first type is called ocean thermal energy. This uses warm water on the surface of the ocean to generate power. The second is ocean

mechanical energy, which uses the force of the tides and waves to create power. It's a consistent form of renewable energy that can fill in where the less-consistent renewable energy sources fail us. One of the most beneficial aspects of ocean energy is its huge potential. For example, one mile's worth of wave crests along a coastline[35] creates enough energy to power about 40,000 homes. One of the disadvantages of ocean energy is that there aren't many locations around the world where people can build tidal barrages — the names of the structures that store ocean energy — to collect that energy. Also, the patterns of the tides dictate that ocean energy can only provide power for around 10 hours each day. Especially intense waves might damage or destroy the tidal barrages as well.	[35] *n.* 海岸线

参 考 译 文

人工智能在电气工程中的应用

1. 人工智能简介

人工智能（AI）是计算机科学的一个领域，专注于开发可以智能化、自主工作的技术。人工智能利用先进的算法使计算机能够执行通常需要人类意识而完成的任务，如做出决策、解决问题和理解环境。它在电气工程中具有广泛的应用，从电力系统自动化到工业机器学习解决方案。本文将探讨人工智能如何在电气工程中应用，并展示当前使用这些技术的项目的一些示例。

2. 人工智能及其用途概述

人工智能是一个快速发展的技术领域，它使机器能够以越来越自主和智能的方式行动。在电气工程中，人工智能驱动的系统用于各种应用，从仿真和建模到机器人设计和流程优化。人工智能可用于自主控制电网和变电站等其他大型电力设施，并提高管理复杂制造流程的效率。它还可用于工业工厂内的预测性维护策略。此外，它能改进智能建筑的能源管理，并通过自动化减少人工错误。总之，人工智能在彻底改变整个电气工程行业方面具有巨大潜力，其广泛的用途使组织能够利用机器学习算法来优化在各个领域的运营。

3. 人工智能在电气工程中的应用

人工智能在包括电气工程在内的许多不同行业中越来越受到关注。它可用于帮助执行自动化能源预测，以及优化电网系统以提高效率等任务。人工智能系统旨在使用现有的基础设施和从传感器获取的数据来构建预测模型并做出决策，从而提高性能、节省成本和更有效地利用资源。人工智能算法还可用于产品开发的自动化，如计算机辅助设计（CAD）、计算机辅助制造（CAM）、大数据集的模拟和分析。此外，电力电子工程师利用被称为赫布网络的专用强化学习架构，该架构在面对不稳定的计算环境时可以高效运行。使用这种架构有助于高压电力线的设计和智能汽车或无人机的开发。

4. 人工智能在电气工程中的益处和挑战

在电气工程中使用人工智能具有许多优势。通过人工智能，工程师可以利用复杂的模型和技术来寻找新的解决方案并优化现有的解决方案。此外，在电气工程中使用机器学习算法可以自动执行许多烦琐的任务，而这些任务由人工完成颇为费力。自动化还带来了潜在的成本节约，因为不同的流程都得到了简化，并且在更短的时间内找到了更高效的设计，从而降低了劳动力成本。

尽管将人工智能纳入电气工程项目具有多种好处，但在广泛采用这项技术之前仍然存在一些挑战。问题之一是数据可用性。尽管传感器技术在提供实时数据集方面取得了进步，但它们仍然非常昂贵且不可靠，这使得预算有限的小型组织很难充分利用人工智能所提供的功能。此外，由于最近的进展集中在训练自主机器上，因此在可解释性方面存在重大障碍。这些机器可能能够解释复杂的数据，但在做出决策时无法产生有意义的解释。这意味着对于那些不了解深度学习算法背后的底层技术的人来说，更难做出准确的决策。因此，领导团队有时不信任通过自动化系统获得的结果。最后，还需要考虑"黑盒"问题，其中输入和输出之间可能存在未被发现的不准确的相关性，但如果这些系统提出的假设被视为事实而不是估计，则可能会导致有害的后果。

5. 人工智能自动化电网管理

人工智能具有彻底改变电气工程领域的巨大潜力。通过使用人工智能技术，电网管理可以实现自动化，以提供高效、可靠的能源供应，并提高安全性。机器学习等人工智能技术通过识别过去数据中可能不会注意到的模式来检测电力系统中的异常情况。由于具有预测分析功能，因此这种自动化可以在涉及与电网状态相关的问题时响应更快。此外，人工智能系统可以更好地预测需求和平衡峰值负载，以确保即使在高使用期也能稳定供电。利用人工智能技术可以实现电网自动化，它通过后台运行的智能算法实现更加智能的决策过程，这样可在宏观层面提高整个国家能源供应的效率和可靠性。

6. 利用人工智能开发智能电气系统

人工智能越来越多地被用于开发各种行业和应用中的智能电气系统。人工智能技术可用于为各种类型的设备创建自动化、自适应和智能控制系统，从控制电动阀门和泵的执行器到控制机器人等机器的数字电机。人工智能算法还可以为变压器或断路器等主要部件提供预测维护能力，以提高系统可靠性。这可以减少停机时间和停机相关成本，无须手动检查且不会牺牲安全标准。此外，它通过提供复杂操作的反馈来帮助行业工程师增加他们的知识库，这些反馈有助于根据长期收集的历史数据点做出预测性决策。最终，使用人工智能可以更有效地利用资源，同时保持高产量，其准确性比传统方法高得多。

7. 人工智能在电气工程自动化质量控制中的应用

由于人工智能可以快速准确地检测和识别复杂模式，因此在电气工程中使用人工智能进行自动化质量控制已变得越来越流行。人工智能驱动的系统可以快速检测电机控制、电路设计和嵌入式系统开发等过程中的数据结构与预期结果之间的差异和不一致。基于人工智能算法对质量控制过程进行准确分析，可以配置最佳目标及其相关参数，这大大减少了测试电气工程设备和电路所需的时间，同时也消除了与手动操作相关的错误。此外，自动化有助于更精确地跟踪合规性测量，从而提高整体产品质量和耐用性。

8. 人工智能在电气工程诊断中的应用

人工智能可以显著提高诊断电气工程设备的准确性和效率。人工智能驱动的检测和电子监控系统可以对复杂的电子设备进行预防性维护。这些系统采用的算法允许系统检测异常；识别环境温度或电压变化等因素，快速解释从传感器或其他数据源检索的数据，并且在工程师的最小干预下准确地计算根本原因分析（RCA）。通过利用人工智能驱动的预测分析，可以制定预防性维护计划，以便在潜在问题出现之前主动预测它们。这不仅确保了最佳的运行性能，还降低了由于时间推移可能出现的不可预见问题而导致的紧急维修成本。

9. 用于电气系统维护的机器学习

机器学习作为一种有效维护电气系统的有益技术正在迅速发展。通过使用机器学习算法可以高精度预测关键组件的故障，使工程师和技术人员能够制定程序化的维护计划，从而节省时间和金钱，并确保系统的最佳效率。这些预测还可以在系统中的潜在薄弱环节造成任何中断或昂贵的维修之前发现问题，从而最大限度地减少停机时间，以及生产损失或环境破坏造成的负面影响。聚类、网络分析和模式识别等机器学习技术特别适合对故障代码和示波器读数等电气数据源进行建模，以便在电气系统的整个生命周期内持续准确地监控电气系统的组件。

10. 结论

由于人工智能具有优化解决方案、自动化任务和解决复杂问题的潜力，因此它在电气工程领域变得越来越流行和应用。人工智能技术能够用于可再生能源项目、智能电网自动化、故障检测和预防系统、基于无人机的检查操作、基于神经网络的电路设计等应用。利用人工智能的公司可能会利用自动化大幅降低成本，并且由于先进的优化能力，仍然可以提供高质量的最终产品。最终，这可能会提高电气工程行业中许多企业的竞争优势。

附录 A 自测题及参考答案

1. 根据英文单词，写出中文意思（20×0.5=10，共 10 分）。

英 文 单 词	中 文 意 思
actuator	
analysis	
circuit	
conversion	
distortion	
gate	
linear	
module	
parameter	
rechargeable	
reverse	
signal	
switch	
vacuum	
workstation	
status	
aging	
breakdown	
coil	
decouple	

2. 根据中文意思，写出英文单词（20×0.5=10，共 10 分）。

中 文 意 思	英 文 单 词
adj. 电气化学的	
n. 发电机	
n. 负荷，负载，加载	
n. 网络	
n. 模式	
n. 冗余	
adj. 健壮的	
n. 硅，硅元素	

中 文 意 思	英 文 单 词
adj. 三相的	
vt. 检验，校验，查证，核实	
n. 向量，矢量	
n. 标准化	
n. 交流发电机	
n. 电容	
n. 电导，导体，电导系数	
n. 损耗	
adj. 电磁的	
n. 安装；装置	
n. 仪表，计，表	
n. 欧姆计，电阻表	

3. 根据英文词组，写出中文意思（15×1=15，共15分）。

英 文 词 组	中 文 意 思
autonomous machine	
digital motor	
neural network	
digital signal processing	
magnetic field	
electric current	
conducting plate	
variable capacitor	
analog circuit	
circuit breaker	
electronic monitoring system	
pattern recognition	
fault analysis	
peak value	
electromotive force	

4. 根据英文缩写，写出英文完整形式及中文意思（10×2=20，共20分）。

	英文完整形式	中 文 意 思
MOSFET		
PCB		
PLC		
RAM		
ROM		
SoC		

	英文完整形式	中文意思
UIC		
ADC		
AI		
CAD		

5. 翻译句子（5×3=15，共 15 分）。

(1) Ammeters typically include a galvanometer; digital ammeters typically include A/D converters as well.

(2) Generally, an amplifier is a device for increasing the power of a signal.

(3) In electromagnetism and electronics, capacitance is the ability of a body to hold an electrical charge.

(4) All conductors contain electric charges which will move when an electric potential difference (measured in volts) is applied across separate points on the material.

(5) A dielectric is an electrical insulator that can be polarized by an applied electric field.

6. 把下列句子翻译成英文（5×2=10，共 10 分）。

（1）典型的电感是由导线绕成的线圈。

（2）电流表是一种用来测量电路中电流的测量仪。

（3）交流发电机是机电设备，它把机械能转换成交流电形式的电能。

（4）电容器是一种被动电子元件，它由一对以电介质（绝缘体）分隔的导体组成。

（5）在像铜或铝这样的金属导体中，可移动的带电粒子就是电子。

7. 根据下列方框中所给的词填空（10×1=10，共 10 分）。

| (A) vacuum tube (B) forward (C) direction (D) rectification (E) semiconductor |
| (F) reverse (H) alternating current (I) cathode (J) signals (K) electronic component |

In electronics, a diode is a two-terminal ___1___ that conducts electric current in only one ___2___. The term usually refers to a semiconductor diode, the most common type today. This is a crystalline piece of ___3___ material connected to two electrical terminals. A vacuum tube diode (now little used except in some high-power technologies) is a ___4___ with two electrodes: a plate and a ___5___.

The most common function of a diode is to allow an electric current to pass in one direction (called the diode's ___6___ direction), while blocking current in the opposite direction (the ___7___ direction). Thus, the diode can be thought of as an electronic version of a check valve. This unidirectional behavior is called ___8___, and is used to convert ___9___ to direct current, and to extract modulation from radio ___10___ in radio receivers.

8. 根据下列短文回答问题，回答请使用英文（5×2=10，共 10 分）。

Electric Motor

An electric motor converts electrical energy into mechanical energy. Most electric motors operate through interacting magnetic fields and current-carrying conductors to generate force. The reverse process, producing electrical energy from mechanical energy, is done by generators such as an alternator or a dynamo. Many types of electric motors can be run as generators and vice versa. For example, a starter/generator for a gas turbine or traction motors used on vehicles often perform both tasks. Electric motors and generators are commonly referred to as electric machines.

Electric motors are found in applications as diverse as industrial fans, blowers and pumps, machine tools, household appliances, power tools, and disk drives. They may be powered by direct current (e.g., a battery powered portable device or motor vehicle), or by alternating current from a central electrical distribution grid. The smallest motors may be found in electric wristwatches. Medium-size motors of highly standardized dimensions and characteristics provide convenient mechanical power for industrial uses. The very largest electric motors are used for propulsion of ships, pipeline compressors, and water pumps with ratings in the millions of watts. Electric motors may be classified by the source of electric power, by their internal construction, by their application, or by the type of motion they give.

The physical principle of production of mechanical force by the interactions of an electric current and a magnetic field was known as early as 1821. Electric motors of increasing efficiency were constructed throughout the 19th century, but commercial exploitation of electric motors on a large scale required efficient electrical generators and electrical distribution networks.

Some devices convert electricity into motion but do not generate usable mechanical power as a primary objective and so are not generally referred to as electric motors. For example, magnetic solenoids and loudspeakers are usually described as actuators and transducers, respectively, instead of motors. On the other hand, some electric motors are simply used as a means of producing torque or force (such as magnetic levitation), but in many cases are still considered to be electric motors, even though they do not actually generate any mechanical energy per se.

Questions:
(1) What does an electric motor do?
(2) How do most electric motors operate to generate force?
(3) What are the very largest electric motors used for?
(4) What did commercial exploitation of electric motors on a large scale require?
(5) What are magnetic solenoids and loudspeakers usually described as respectively?

参 考 答 案

1. 根据英文单词，写出中文意思（20×0.5=10，共 10 分）。

英 文 单 词	中 文 意 思
actuator	*n.* 执行器，执行机构
analysis	*n.* 分析，分解
circuit	*n.* 电路
conversion	*n.* 转换，变换
distortion	*n.* 变形，失真
gate	*n.* 逻辑门
linear	*adj.* 线的，直线的，线性的
module	*n.* 模块；组件
parameter	*n.* 参数，参量
rechargeable	*adj.* 可再充电的
rever	*n.* 转子
signal	*n.* 信号
switch	*n.* 开关，电闸；转换
vacuum	*n.* 真空
workstation	*n.* 工作站
status	*n.* 状态
aging	*n.* 老化
breakdown	*n.* 损坏，故障
coil	*n.* 线圈，绕组
decouple	*vt.* 去耦，解耦

2. 根据中文意思，写出英文单词（20×0.5=10，共 10 分）。

中 文 意 思	英 文 单 词
adj. 电气化学的	electrochemical
n. 发电机	generator
n. 负荷，负载，加载	load
n. 网络	network
n. 模式	pattern
n. 冗余	redundancy
adj. 健壮的	robust
n. 硅，硅元素	silicon
adj. 三相的	three-phase
vt. 检验，校验，查证，核实	verify

中 文 意 思	英 文 单 词
n. 向量，矢量	vector
n. 标准化	standardization
n. 交流发电机	alternator
n. 电容	capacitance
n. 电导，导体，电导系数	conductance
n. 损耗	depletion
adj. 电磁的	electromagnetic
n. 安装；装置	installation
n. 仪表，计，表	meter
n. 欧姆计，电阻表	ohmmeter

3. 根据英文词组，写出中文意思（15×1=15，共 15 分）。

英 文 词 组	中 文 意 思
autonomous machine	自动机器
digital motor	数字电动机，数字马达
neural network	神经网络
digital signal processing	数字信号处理
magnetic field	磁场
electric current	电流
conducting plate	导电板
variable capacitor	可变电容器
analog circuit	模拟电路
circuit breaker	断路开关，断路器
electronic monitoring system	电子监控系统
pattern recognition	模式识别
fault analysis	事故分析，故障分析
peak value	峰值
electromotive force	电动势

4. 根据英文缩写，写出英文完整形式及中文意思（10×2=20，共 20 分）。

	英文完整形式	中 文 意 思
MOSFET	Metallic Oxide Semiconductor Field Effect Transistor	金属-氧化物-半导体场效应晶体管
PCB	Printed Circuit Board	印制电路板
PLC	Programmable Logic Controller	可编程逻辑控制器
RAM	Random Access Memory	随机存储器
ROM	Read Only Memory	只读存储器
SoC	System on Chip	单片系统

	英文完整形式	中文意思
UIC	Use Initial Conditions	使用初始条件
ADC	Analog to Digital Converter	模数转换器
AI	Artificial Intelligence	人工智能
CAD	Computer Aided Design	计算机辅助设计

5. 翻译句子（5×3=15，共 15 分）。

（1）电流表通常包括一个电流计，而数字电流表通常还包括 A/D 转换器。

（2）一般来说，放大器是一个提高信号功率的器件。

（3）在电磁学和电子学中，电容是能够保存电量的一个部件。

（4）所有导体都含有电荷，当该材料不同端点间存在电位差时（用伏特计测量）电荷就会移动。

（5）电介质是电绝缘体，可以通过外加电场极化。

6. 把下列句子翻译成中文（5×2=10，共 10 分）。

(1) Typically an inductor is a conducting wire shaped as a coil.

(2) An ammeter is a measuring instrument used to measure the electric current in a circuit.

(3) An alternator is an electromechanical device that converts mechanical energy to electrical energy in the form of alternating current.

(4) A capacitor is a passive electronic component consisting of a pair of conductors separated by a dielectric (insulator).

(5) In metallic conductors, such as copper or aluminum, the movable charged particles are electrons.

7. 根据下列方框中所给的词填空（10×1=10，共 10 分）。

(1) K　(2) C　(3) E　(4) A　(5) I　(6) B　(7) F　(8) D　(9) H　(10) J

8. 根据下列短文回答问题，回答请使用英文（5×2=10，共 10 分）。

(1) An electric motor converts electrical energy into mechanical energy.

(2) Most electric motors operate through interacting magnetic fields and current-carrying conductors to generate force.

(3) The very largest electric motors are used for propulsion of ships, pipeline compressors, and water pumps with ratings in the millions of watts.

(4) Commercial exploitation of electric motors on a large scale required efficient electrical generators and electrical distribution networks.

(5) Magnetic solenoids and loudspeakers are usually described as actuators and transducers, respectively.